Recent Trends in
INVERTEBRATES

Vol. 8

H.S.Bhamrah
Kavita Juneja

ANMOL PUBLICATIONS

ANMOL PUBLICATIONS
4378/4B, Ansari Road,
Darya Ganj, New Delhi - 110002.

Recent Trends in Invertebrates

ISBN 81 7041 473 3 (Set)
ISBN 81 7041 474 1 (Vol.I)
ISBN 81 7041 475 X (Vol.II)
ISBN 81 7041 476 8 (Vol. III)
ISBN 81 7041 477 6 (Vol. IV)
ISBN 81 7041 478 4 (Vol. V)
ISBN 81 7041 479 2 (Vol. VI)
ISBN 81 7041 480 6 (Vol. VII)
ISBN 81 7041 481 4 (V01. VIII)

PRINTED IN INDIA

Published by J.L. Kumar for Anmol Publications and Laser Typeset by D.V. Computers, Delhi, Printed at Efficient Offset Printers, New Delhi.

PREFACE

The present title has been designed to meet the requirements of the undergraduate students. In the Recent Trends in Invertebrate all phyla have been discussed. The representative types of each phylum have been described in detail. This is followed by the classification of the phylum with a description of its common animals. Recent findings and work have been incorporated in the text. Special care has been taken to make the language as simple and lucid as possible. Simple and accurate diagrams accompany the text. In the interest of the students efforts have been made to cover the syllabi of nearly all Indian Universities.

The book features both a text and a laboratory guide. It deals with the structures and physiological phenomenon of each system. General topics are also dealt in details. It is hoped that this book will not only meet the requirement of undergraduate students but also be useful as a guideline to the teachers in their teaching. It is also hoped that the general public will find that the book is not riddled with too much of scientific and technical jargon.

It is difficult to acknowledge adequately the assistance of all who have contributed to the preparation of this book. First of all the authors express their thanks to their friends and colleagues whose continuous inspiration have initiated them to bring out the present volume.

The authors are grateful to Dr. J. Bahadur, Professor and Head, School of Studies in Zoology, Jiwaji University Gwalior, Dr. A.K. Pandey, Professor and Head, School of Studied in Zoology, Vikaram University, Ujjain for their valuable suggestions.

Constructure criticism from students and teachers will be highly appreciated by the authors.

H.S. Bhamrah

Kavita Juneja

Contents

1

CHARACTER AND CLASSIFICATION OF ECHINODERMATA.

1. Small to large animals generally found in sea water.
2. Body colour often brilliant and of different shades, few are almost transparent.
3. Body shape star-like, discoidal, circular or cylindrical with well differentiated oral and aboral surfaces.
4. Triploblastic, coelomate animal without well marked cephalization and segmentation.
5. Radial symmetrical in adult condition but the larva are bilaterally symmetrical.
6. The body wall is made up of three well defined layers, an outer epidermis, middle dermis and innermost peritoneum.
7. Endoskeleton in the form of calcareous, mesodermal plates and sometimes as spines hence the name echinodermata. (Gr., *echinus* = spiny; *derma* = skin).
8. An extensive body cavity or coelom surrounding the internal organs is present.
9. A part of embryonic coelom becomes modified into characteristic *ambulacral* or *water vascular system* or *hydrocoel* with many tube feet. This system is responsible for respiration and locomotion in animal.

10. Alimentary tract is usually coiled tube extending from the mouth located on the oral surface to the anus on the aboral or oral surface.

11. Circulatory or haemal or blood lacunar system is typically present.

12. Respiration occurs through a variety of structures, e.g. papulae in starfishes, peristomial gills in sea urchins, gential bursae in brittle stars and cloacal respiratory trees in holothurians.

13. Excretory system is wanting.

14. Nervous system is primitive, consisting of networks concentrated into the radial ganglionated nerve cords.

15. Sense organs are poorly developed.

16. Sexes are usually separate (dioecious) with few exceptions. Gonads are simple with or without simple ducts.

17. Reproduction is usually sexual, few reproduce asexually or by regeneration.

18. Fertilization is external, while few echinoderms are viviparous.

19. Development is inderterminate including characteristic larvae which undergo metamorphosis into the radially symmetrical adults.

Phylum Echinodermata has been divided into two subphyla;

I. Subphylum - Pelmatozoa

II. Subphylum - Eleuthrozoa

SUBPHYLUM. I. PELMATOZOA

1. It includes both extinct as well as living echinoderms usually attached by a stalk i.e. sedentary.

2. Entire structure pentamerous.

3. Mouth and anus on the oral surface.

4. Tube feet are absent and the tubular ciliated appendages when present are without suckers.

5. Theca is differentiated into aboral cup, calyx and oral roof.
6. Arms are movable with or without pinnules.
7. Main nervous system is aboral. Subphylum Pelmatozoa is differentiated into 5 classes:

Class 1 - Heterostelea

1. Extinct echinoderms with laterally flatenned non-porous theca attached horizontally by a stalk.
2. Mostly without arms extensions.

Example. *Trochocystites.*

Class 2. Blastoidea

1. Extinct echinoderms. Test bud-like, short stemmed or stemless, of 13 major plates.
2. Theca is pentamerous, radial and made up of 13 plates arranged in three rows.
3. Ambulacra five.
4. Characteristic respiratory organs, the *hydrospires,* situated underneath the ambulacra.

Example : *Pentremites.*

Class 3 Cystoidea

1. Extinct animals of oval form attached directly or by a stalk in upright direction.
2. Theca of rigid polygonal plates, of which some are porous.
3. Ambulacra vary from 2-5.

Example : *Crystoblastus*

Class - 4 -Edrioasteroidea

1. Extinct, disc-shaped without stem or arms lying free or attached directly by the aboral surface.

2. Theca cushion-shaped with a star on top formed by five recumbent ambulacral grooves.

3. Anus and madreporite are situated on the oral surface.

Examples : *Stromatocystites, Lepidodiscus*

Class 5 Crinoidea

1. It includes with living as well as non-living echinoderms usually attached by a stalk, but living members are stalkless and free living.

2. Entire structure strongly pentamerous.

3. Body consists of an aboral cup, the calyx and oral cover or roof, the *tegmen.*

4. Oral surface directed upwards.

5. Mouth usually central, anus usually excentric are present on the oral surface.

6. Arms movable, simple, mostly branched, usually five or ten in number with or without pinnules.

7. Ambulacral groves are open and extend along arms and pinnules to their tips.

8. Madreporite, spines and pedicellariae are present.

9. Sexes are separate. Development indirect through doliolaria larva.

Class Crenoidea is divided into four orders:

Order (i) Articulata

1. Extinct as well as living forms.

2. Calyx pentamerous flexible and dicyclic.

3. Mouth and ambulaceral grooves are not covered by tegminal plates and remain exposed.

4. Tegman leathery and contains small calcareous plates or minute ossicles.

Examples : *Metacrinus, Rhizocrinus, Antedon, Comatula.*

Order (ii) Inadunata

1. Extinct forms.
2. Calyx rigid and lower arm ossicles spring free from it.
3. Mouth covered over by tegminal plates but ambulacral grooves mostly exposed.
4. Arms three or more with or without pinnules.

Examples : *Anartiocrimus, Hybocystites.*

Order (iii) Flexibilia

1. Extinct forms.
2. Calyx flexible, dicyclic.
3. Arms branched, devoid of pinnules and typically curved inwards.
4. Mouth and ambulacra exposed.

Example : *Forbesiocrinus.*

Order (iv) Camerata

1. Extinct forms.
2. Calyx rigid, mono - or dicyclic.
3. Arms branched and pinnulate.
4. Mouth and ambulacra are covered.

Example : *Xenocrinus.*

SUBPHYLUM II. ELEUTHROZOA

1. Mostly living echinoderms.
2. Stem or stalk absent, usually free-living forms.
3. Body structure usually pentamerous.
4. Oral surface bearing the mouth is downward or lying on one side.

5. Anus usually on the aboral surface.
6. Ambulacral grooves usually not for food gathering and the tube feet with suckers are chiefly locomotory organs.
7. Main nervous system is oral.

Subphylum Eluthrozoa is divided into four classes.

1. Class - Asteroidea
2. Class - Ophiuroidea
3. Class - Echinoidea
4. Class - Holothuroidea

Class 1. Asteroidea

1. Body flattened and star-shaped with a central diac and five radiating arms which are not sharply marked off from the disc.
2. Arms hollow and contain gonads, prolongation of gut, coelom and other visceral organs.
3. Ambulacral grooves open in tube feet and extend upto the tip of arms.
4. Madreporite and pedicellaria sessile.

The modern classification distinguishes class Asteroidea into two fossil and three living orders :

Order (i) Phanerozonia

1. Arms short with broad bases and with two rows of large marginal plates.
2. Papullae limited to the aboral surface.
3. Pedicellariae generally sessile or absent.
4. Tube feet with or without suckers.

Exmaples: *Luidia, Oreaster, Pentaceros, Astropecten* .

Order (ii) Spinulosa

1. Arms generally without conspicuous marginal plates.

2. Pedicillariae rare.
3. Podia or tube feet are in two rows provided with suckers.
4. Aboral surface with low spines.
5. Ampullae single or bifurcated.

Examples : *Asterina, Solaster, Hymenaster.*

Order (iii) Forcipulata

1. No conspicuous marginal plates.
2. Aboral skeleton mostly reticulate with conspicuous spines.
3. Pedicillariae pedunculate and straight or crossed type.
4. Tube feet arranged in four rows provided with suckers.
5. Papulae on both surface.

Example : *Asterias, Heliaster, Brisingaster.*

Order (iv) Platyasterida

1. Extinct forms.
2. Ambulacra widely open. Arms with series of long interambulacral ossicles.

Example : *Platasterias.*

Order(v) Hemizonida

1. Extinct forms.
2. Ambulaceral grooves are deep and ambulacral ossicles are in-sunk.

Example : *Palasterina, Helianthaster.*

Class 2. Ophiuroidea

1. Marine usually deep-sea and free living echinoderms.
2. Body flaterred with a pentamerous or rounded central disc.
3. Oral and aboral surfaces are distinct.

4. Arms usually five slender, jointed, solid, muscular, flexible rarely six or seven.
5. No ambulacral grooves.
6. Pedicillariae, skill gills and special sense organs are absent.
7. Madreporite opens on the oral surface.
8. Anus and alimentary canal are absent.
9. Tube feet arising ventrolaterally from arms in 2 rows. Ampullae and suckers are absent.
10. Sexes are separate, gonads pentamerous.
11. Development includes a free-swimming *Pluteus larva.*

Class Opiuroidea is divided into two orders:

Order (i) Ophiurae

1. Arms simple, mostly five in number, moving chiefly in transverse plane.
2. Arm ossicles articulated by pits and projections.
3. Disc and arms are usually covered with distinct shields or scales.
4. Arms spines are borne laterally and are directed outward or toward the arm tips, not downwards.
5. Single madreporite.

Examples: *Ophioderma, Ophioscolex, Ophiothrix, Ophiocoma.*

Order (ii) Euryalae

1. Arms simple or branched.
2. Scalation of arms and disc poorly developed.
3. Arms move vertically and may be entwined around the objects.
4. It includes *Brittle-stars*.

Example : *Astrophytòn*.

Class -3 Echinoidea

1. Body spherical disc-like, oval or heart shaped.
2. The body is lying enclosed in a shell formed of by closely fitting calcareous plates.
3. Outer calcareous plates are distinguished into five alternating ambulacral and five inter-ambulacral areas.
4. The oral and abroal surfaces are distinct, the mouth is located on oral surface while the anus and madreporite are present on the aboral side.
5. Tube feet come out from the pores of ambulacral plates and are locomotory organelles.
6. Pedicillariae are stalked and three jawed.
7. Sexes separate and, glanads 5 in number and interradial.
8. Development includes free-swimming *echinopluteus larva.*

Class Echinoidea is divided in three subclasses:

Sublass (A) Bothriocidaroida

1. A single row of plates in each interambulacral area.
2. Madreporite radial.
3. Lantern of Aristotle absent.

Example : *Bothriocidaris* (extinct).

Subclass (B) Regularia or Endocyclica

1. It includes globular, circular and oval forms, which are pentamerously symmetrical with two rows of inter-ambulacral plates.
2. Mouth and anus lie at opposite pole.
3. An Aristotle's lantern is present.
4. Madreporite is ambulacral.

Order (i) Lepidocentroida

1. Test flexible with overlapping or separated plates.

2. Ambulacral plates continue upto mouth lips.

Examples : *Palaeodiscus, Sperosoma.*

Order (ii) Melonechinoida

1. Test spherical, rigid.
2. Ambulacral plates continue upto mouth lip.
3. Interambulacral plates in 4 or more rows.

Example : *Melonechinus.*

Order (iii) Cidaroidea.

1. Test rigid and globular.
2. Two rows of long narrow ambulacral plates and two rows of inter-ambulacral plates present.
3. Ambulacral and inter-ambulacral plates continue up to mouth lip.
4. Gills and sphaeridia are absent.
5. Five bushy Stewart's organs are present appended to the lantern.

Examples : *Cidaris, Notocidaris.*

Order (iv) Centrechinoida

1. Test rigid and rarely oval.
2. Peristome with 10 buccal plates.
3. Anus surrounded by a double circle of apical plates.
4. External gills and sphaeridia present.

Examples: *Echinus, Solenia, Diadema, Arbacia.*

Subclass - (C)- Irragularia or Exocyclica

1. Bilaterally symmetrical. Test mostly flaterred oval or circular.
2. Mouth centrally placed on the oral surface.
3. Anus is displaced posteriorly generally marginal at oral or aboral surface and lies outside the apical system of plates.

4. Podia or tube feet not locomotory.

Order (i) Holectypoida

1. Test regular with simple ambulacra and centrally located peristome and apical system.
2. Ambulacra narrow on oral surface.
3. Lantern teeth with lateral flanges

Example : *Holectypus* (Extinct).

Order (ii) Cassiduloida

1. Aboral ambulacral area petaloid, forming a 5- armed figure.
2. Lantern absent

Example : *Cassidulus* (Extinct)

Order (iii) Clypeastroida

1. Test flattened, oval or rounded in shape covered with small spines.
2. Mouth and apical system are usually central and oral in position.
3. Aboral ambulacral areas petaloid.
4. Aristotle's lantern present.
5. Gills present.

Exmaples : *Clypeaster, Laganum.*

Order (iv) Spatangoida

1. Test oval or heart-shaped with excentric mouth and anus.
2. Four aboral ambulacral areas petaloid, fifth not petaloid.
3. Lantern absent.
4. Gills absent.

Examples : *Spatangus, Lovenia, Hemiaster.*

Class. 4. Holothuroidea

1. Body bilaterally symmetrical, usually enlongated in the oral-aboral axis having mouth at near one end and anus at or near the other end.
2. Endoskeleton reduced to microscopic spicules or plates embedded in the body wall.
3. Mouth surrounded by a set of tentacles attached to water vascular system.
4. Podia or tube feet are usually present and locomotory.
5. Alimentary canal is long and coiled and cloaca usually with respiratory trees.
6. Sexes are usually separate and gonad single or made of one or two tufts of tubules.
7. Development may be direct or indirect with a larva called *auricularia*

Order (i) Aspidochirota

1. Podia or tube feet numerous.
2. Mouth is surrounded by 10-30 mostly 20 peltate or branched oral tentacles.
3. Retractor muscles of pharynx absent.
4. A pair of well developed respiratory trees is present.

Examples : *Holothuria, Stichopus, Mesothuria.*

Order (ii) Dendrochirota

1. Holothurians with numerous podia.
2. Oral tentacles dendroid.
3. Retractor muscles and respiratory tree present.
4. Madreporite internal.

Examples : *Cucumaria, Thyone,* etc,

Order (iii) Elasipoda

1. Oral tentacles 10-20, peltate or shield shaped.
2. Tentacular ampullae, oral retractors and respiratory trees are absent.
3. Madreporite may be internal or external.
4. Body generally flat ventrally, with mouth usually ventral.
5. Tube feet few.

Examples : *Elpidia, Pelagothuria.*

Order (iv) Malpadonia

1. Oral tentacles unbranched and digitate or with minute branches at the tip.
2. Madreporite internal and respiratory tree present.
3. Tube feet absent and posterior end of body tapers to form caudal region.

Examples: *Malpodia, Caudina.*

Order (v) Synaptida or Apoda

1. Body vermiform having smooth or warty surface.
2. Podia or tube feet absent.
3. Oral tentacles are 10-20. Simple, digitate or pinnate.
4. Pharyngeal retractors are present in some forms.
5. Respiratory trees are absent.
6. Water vascular system greatly reduced.

Examples : *Synapta, Chiridota*

METACRINUS

Phylum	-	Echinodermata
Subphylum	-	Pelmatozoa
Class	-	Crinoidea

Order - Articulata
Genus - *Metacrinus*

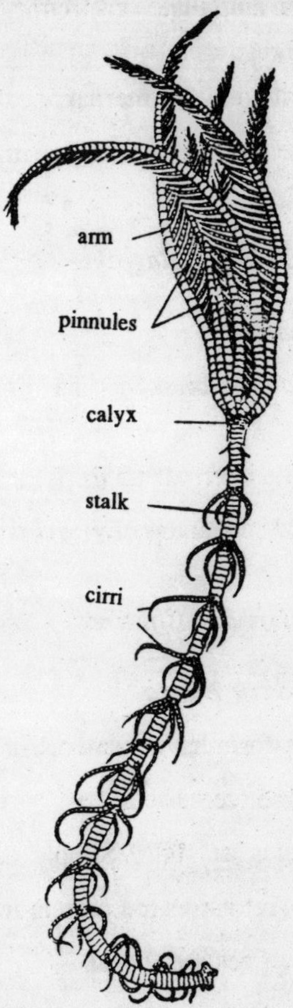

Fig. 1.1. *Metacrinus*

1. It is commonly known as 'sea-lily'.
2. It is a marine animal found at moderate depths. It is found in Japan, Malaya and Australian region.
3. The calyx is small and appears to be monocyclic

4. It has four branchials between the radials and first branching.
5. The five arms are long, each dichotomouldy divided three times and with small uniserial pinnules.
6. Stalk is long, distinctly pentamerous and jointed.
7. Around the joints are present whorls of numerous, slender, many jointed cirri, which are used for temporary attachment.

CTENODISCUS

Phylum	-	Echinodermata
subphylum	-	Eleutherozoa
Class	-	Asteroidea
Order	-	Phanerozonia
Genus	-	*Ctenodiscus*

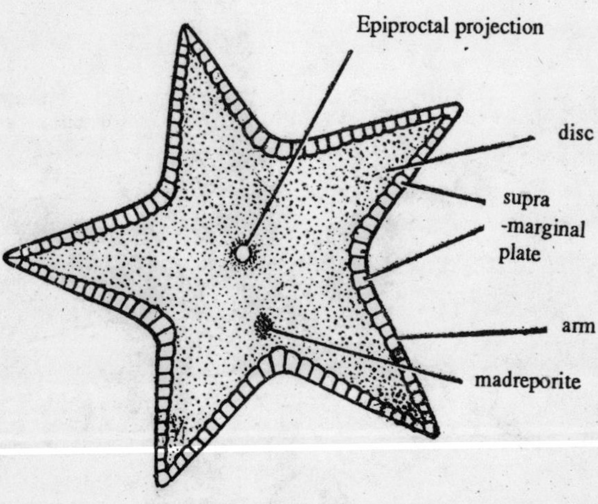

Fig. 1.2. *Ctenodiscus*

1. It is commonly known as 'mud-star' found in a shallow water circumpolar type ranging from cold northern waters to California, Japan and Cape cod.

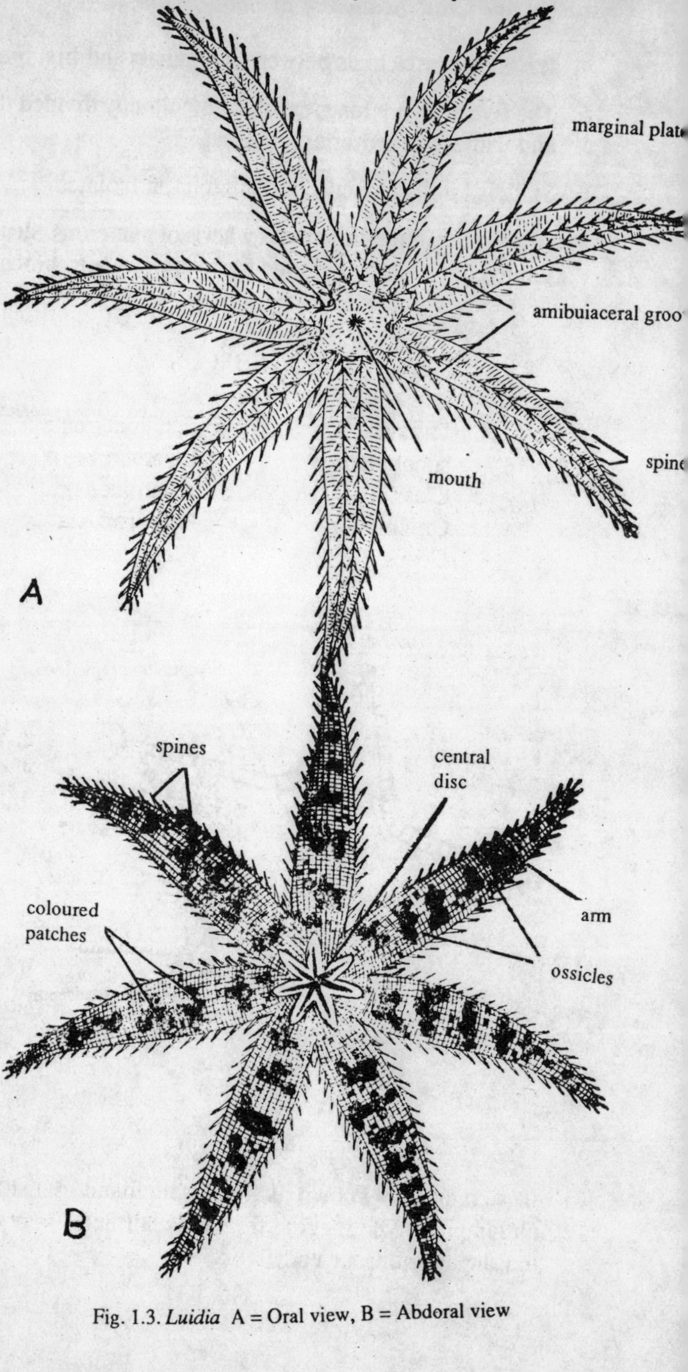

Fig. 1.3. *Luidia* A = Oral view, B = Abdoral view

2. Body from varies form pentagonal to stellate with conspicuous marginal plates without spines.
3. The aboral wall is thin, membranous and covered by paxillae and a tube-like epiproctal projection arises from its centre.
4. Between certain adjacent marginal plates occur vertical rows of thin folds of integument, called cribiform organ.
5. Intestine, intestinal caeca and anus are lacking.

LUIDIA

The systematic position is the same as that of *Ctenodisacus*.

1. It is a littoral form, found in tropical and subtropical seas.
2. Disc is comparatively small and arms are long and flexible.
3. Arms may vary from 5-11 and are bordered with spines.
4. Body is differentiated into oral and aboral surfaces.
5. The oral surface contains a central mouth which communicates with ambulacral groove in each arm.
6. Inframarginal plate is separated from the ambulacral plate by a small intermediate plate.
7. Aboral surface is covered by several small rectangular ossicles and is marked by coloured patches.
8. Anus and intestinal caeca are absent.
9. Pedicillariae are present.
10. Gills are branched.
11. Each ambulacral groove accomodates two rows of tube-feet.
12. A polynoid polychaete (*Achloe astericola*) is found as commensal in the ambulacral groove of *Luidia*.

ASTROPECTEN

The systematic position is same as that of *Ctenodiscus*.

1. It is a large starfish, commonly found in sea below the tide line to great depths.

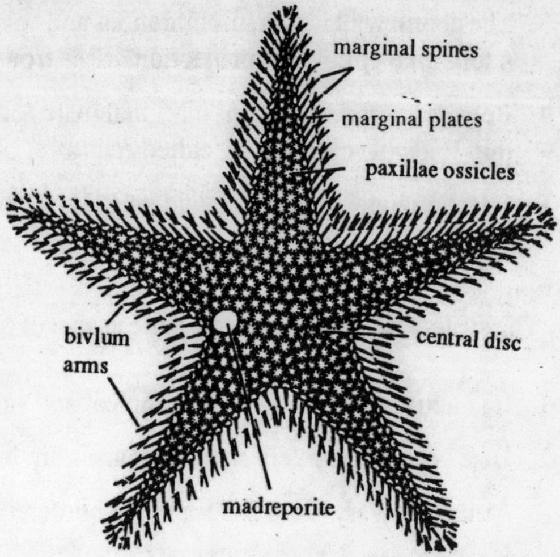

Fig. 1.4. *Astropecten.*

2. It is found in all parts of world namely New Jersey southwards, and California.
3. Body flattened, star shaped with five long slender arms radiating from central disc.
4. Oral and aboral surfaces are distinctly marked.
5. The oral surface contains centrally placed mouth which communicates with 5 broadly opened ambulacral grooves having double rows of tube feet.
6. Tube feet without suckers.
7. The aboral surface has madreporite in one of its interradii.
8. Anus and pedicillariae absent.
9. The stone canal is extremly complicated due to internal development of ridge.
10. The optic cushion contains retinal cells.
11. Sexes are separate. The development includes *brachiolaria larva.*

OREASTER (PENTACEROS)

The systematic position is the same as that of *Ctenodiscus*.

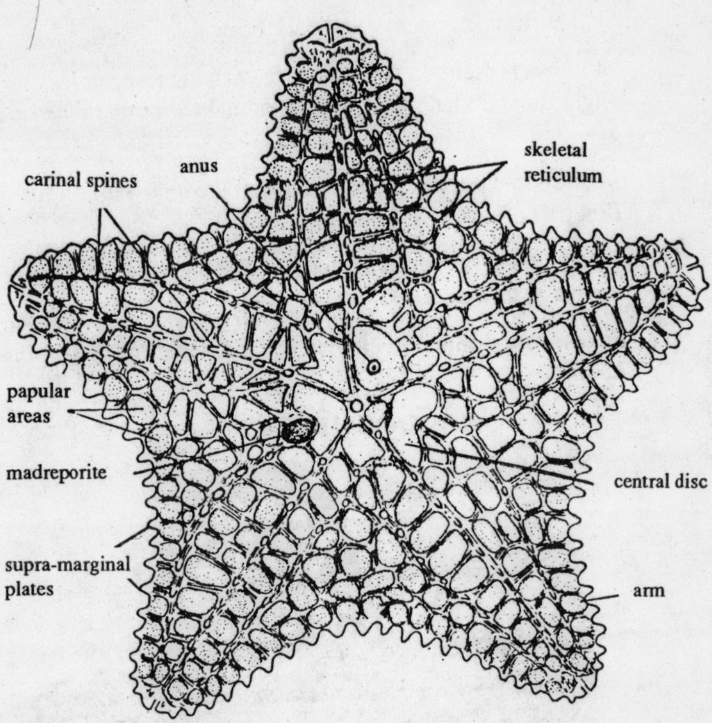

Fig. 1.5. *Oreastes (Pentaceros)*

1. It is most common starfish found in sea from shallow water to 1000 fathoms.
2. It is found in U.S.A., U.K., India and South America.
3. It is commonly called 'sea pentagon'
4. Body is enclosed in hard leathery integument which contains ossicles.
5. Central disc and arms are fused together.

6. Arms are five is number and arranged around a central disc in the form of star.
7. Body surface can be differentiated into oral and aboral surface.
8. Aboral surface is convex and brown in colour.
9. Oral surface is concave and dark brown in colour.
10. Mouth is present on the oral surface and is five angles and each angle is continued into the ambulacral groove up to the tip of an arm.
11. Madreporite plate is present in one of the inter-radii. It forms the first part of water vascular system.
12. The ambulacral grooves contain double row of locomotory organs or tube-feet.
13. Sexes are separate. Development is indirect through a *bipinnaria larva.*

ANTHENEA

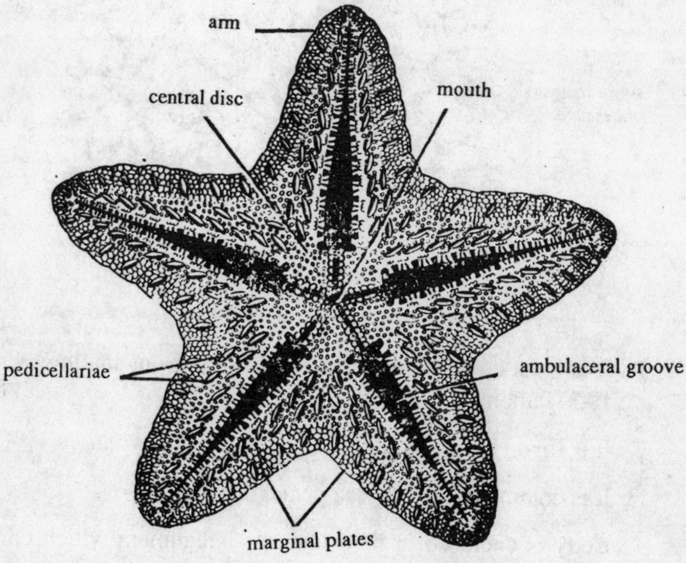

Fig. 1.6. *Anthenea*. Oral View

The Systematic positition is the same as that of *Ctenodiscus.*

1. It is found in all the seas from shallow to deep waters.
2. It is most common in Australian sea.
3. Body consists of relatively large central disc and five relatively short arms.
4. Body is differentiated into oral and aboral surfaces.
5. Aboral surface is convex, having interradial depression.
6. The oral surface is broad and flat containing a central mouth and five ambulacral grooves.
7. The flat surface between the ambulacral groove contains plate-like ossicles beset with a number of minute rounded tubercles and granules which some time assume the shape of spine.
8. Pedicillariae present on both the surfaces.
9. It possesses vertical calcareous and interradial partitions.

SOLASTER

Phylum	-	Echinodermata
Subphylum	-	Eleutherozoa
Class	-	Asteroidea
Order	-	Spinulosa
Genus	-	*Solaster*

1. It is found in Northern waters.
2. Body has 9-14 arms with two rows of marginal plates.
3. The aboral skeleton is reticulate, the plates forming a fine network with small meshes and carrying on a projecting tubercle, a crowded bundle of spinelets.
4. Pedicillariae are absent.
5. Carnivorous, feed on other echinoderms, particularly asteroids.
6. Anus is distinct.

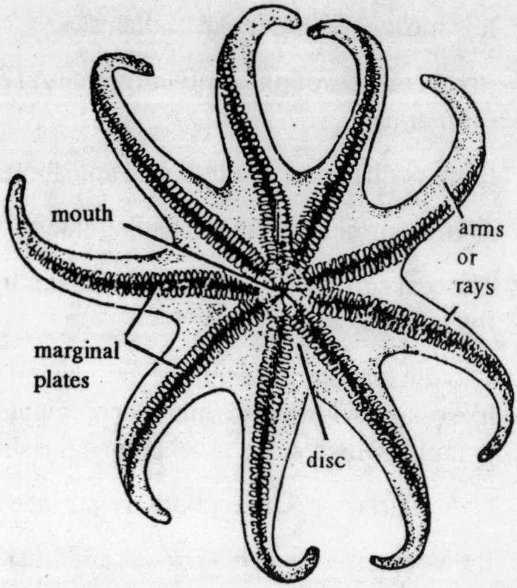

Fig. 1.7. *Solaster*

HELIASTER

Phylum	-	Echimodermata.
Subphylum	-	Eleutherozoa
Class	-	Asteroidea
Order	-	Forcipulata
Genus	-	*Heliaster*

1. It is found in shallow waters in panamic region.
2. It has a broad disc with 40 or more short, tapering arms.
3. Skeleton is reticulate having both straight and crossed pedicillariae.
4. Madreporite is present.
5. Tube feet arise in four rows.
6. Interbranchial septa are double.
7. It begins its life with five rays, but additional rays are added in interradii except that which contains the madreporite.

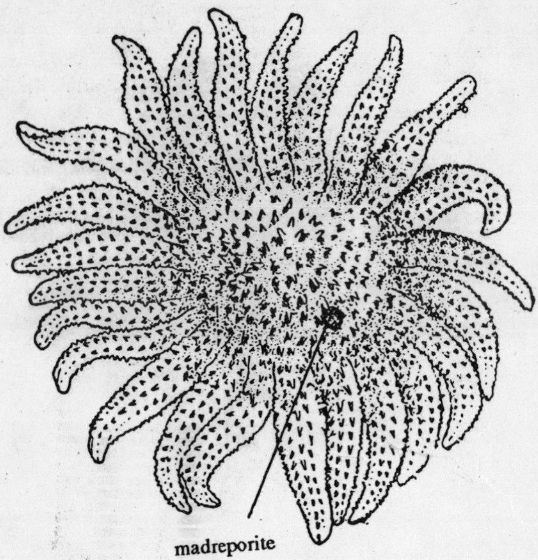

Fig. 1.8. *Heliaster.*

OPHIODERMA

Phylum	-	Echinodermata
Subphylum	-	Eleutherozoa
Class	-	Ophiuroidea
Order	-	Zygophiural
Genus	-	*Ophioderma*

1. *Ophioderma* is a brittle star of littoral zone tropical West Indian seas.
2. Found along the North Atlantic Mediterranean, littoral, tropical and West Indian seas.
3. Body with prominent pentagonal disc of brownish colour having five arms which are sharply distinct from disc and freely movable upon central disc.
4. Arms are slender, jointed, solid, cylindrical, muscular and flexible.

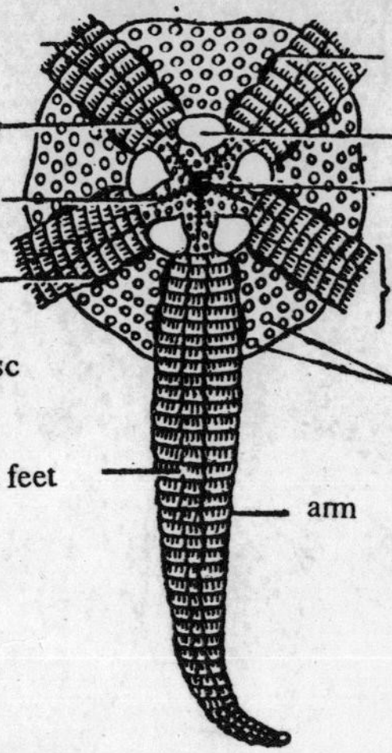

Fig. 1.9. *Ophioderma.* Disc and one arm in oral view.

5. Disc coated with granules.
6. Mouth fringed with numerous oral papillae.
7. Dental papillal are absent and teeth are arranged in a single row.
8. There are four bursal slits in each interradius, two located orally and two peripherally.
9. The spines of arms are very small so that the arms are smooth.
10. Sexes are separate. Development with free-swimming *pluteus larva.*

OPHIOGLYPHA OR OPHIURA

The systematic position is the same as that of *Ophioderma.*

1. *Ophioglypha* is a cosmopolitan brittle star found in almost all sea water.
2. It has cosmopolitan distribution.
3. Body with flat, pentagonal, central disc, covered with radial shields and numerous small scales.

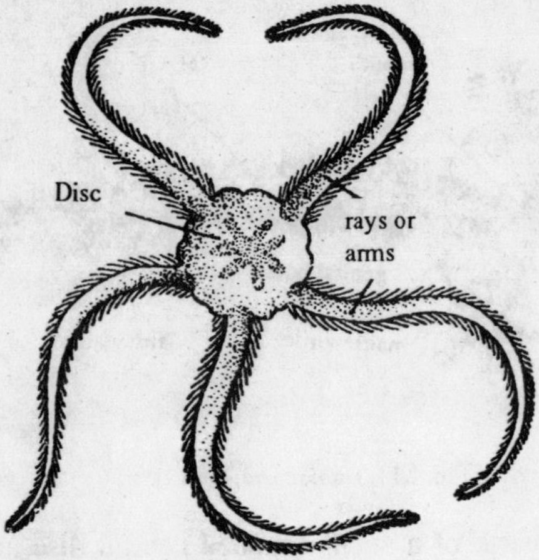

Fig. 1.10. *Ophioglypha*

4. Arms are five and markedly set off from central disc, which are of moderate length and covered with regular plates.
5. The arm plates of lateral side have closely lying spines.
6. Buccal podia of second pair open not into the mouth angles but on the oral surface.
7. Arms combs are present.
8. Bursal slits are two in each inter-radius.
9. The prey is caught by coiling around the arms.
10. Male and female are separate development indirect with *ophiopluteus larva.*

OPHIOCOMA

The systematic position of *Ophiocoma* is same as that of *Ophiouerma.*

1. It is commonly calles 'brittle star'. It is commonly found in shallow waters.
2. It is found along the Red sea shores, West Indian region and Florida.

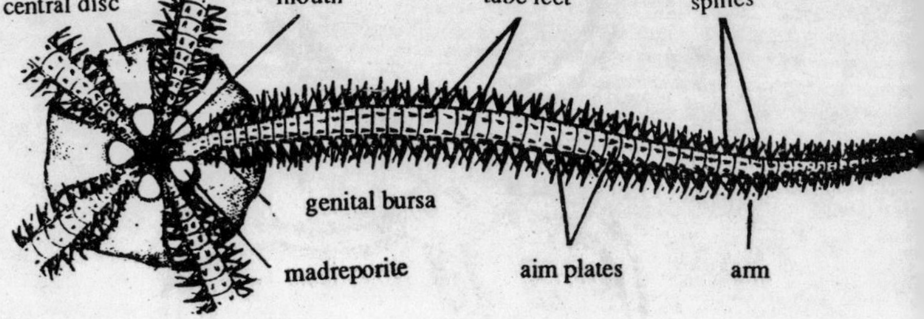

Fig. 1.11. *Ophiocoma*. Oral view.

3. Body consists of a small rounded central disc and five long, slender tapering arms.
4. Oral and aboral surfaces are well marked.
5. Mouth is pentagonal situated in the centre. Madreporite is present in one of the interradial areas.
6. The oral surface has leathery skin.
7. Anus and pedicillarial absent.
8. The ambulacral grooves covered.
9. There are 10 élongated slits, two an either side of the base of each arm called as *genital bursae* in which gonoducts open.
10. Three rows of pointed spines are borne on the lateral plates. A row of short tube feet project on each side between lateral and lower plates.
11. The tube feet are devoid of suckers they are only respiratory and sensory

ASTROPHYTON

Phylum	-	Echinodermata
Subphylum	-	Eleutherozoa

Class - Ophiuroidea
Genus - *Astrophyton.*

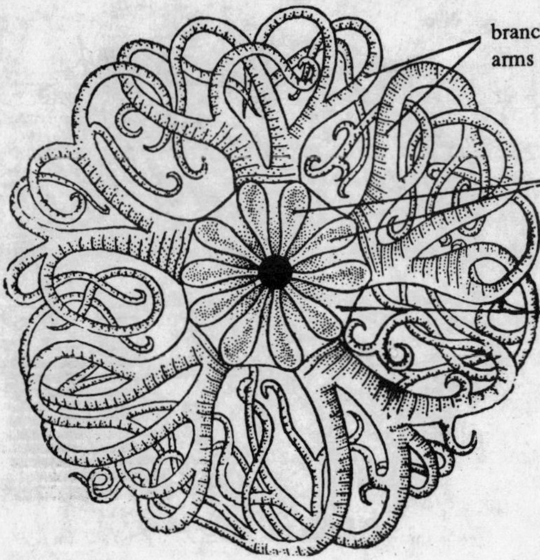

Fig. 1.12. *Astrophyton.*

1. It is commonly known as 'basket star'.
2. It is found in deep sea often rotting on sea bottom.
3. Body surface is soft and devoid of spines.
4. Pentagonal disc is very large and circular.
5. Abroal surface bears spoke-like radial shields which radiate from the centre to the periphery.
6. Fine arms are long, slender and repeatedly branched with tendrillar tip which assist in clinging the objects.
7. One madreporite is present in each inter-radius.
8. Development occurs passing through *ophiopluteus* larva.

GORGONOCEPHALUS

The systematic position is the same as that of *Astrophyton.*

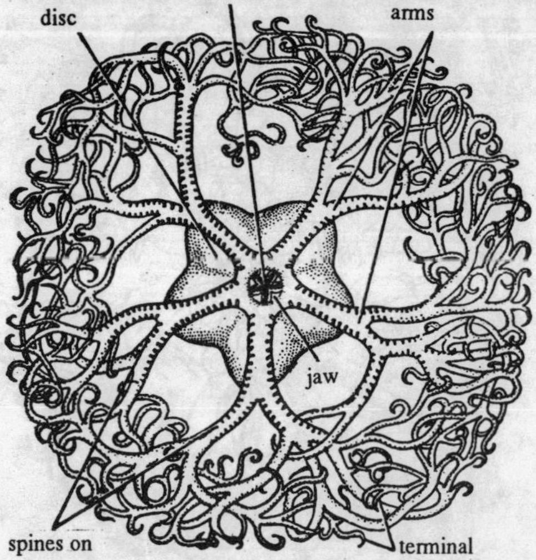

Fig. 1.13. *Gorganocephalus*. Oral View.

1. It is another deep sea water echinoderm and commenly called 'basket star'.
2. The disc is either pentagonal or circular and large.
3. The aboral surface of disc shows long and spoke-like radial shields, radiating from the centre of the disc to the margin.
4. Disc margin between the radial shields bear a series of plates.
5. The five arms are long, slender and flexible and repeatedly branching dichotomously.
6. One medreporite is present is each inter-radius.
7. Bursal slits are short and fusion of bursae from large spaces so that true coelom is greatly reduced.
8. Development occurs through *ophiopluteus* larva.

CLYPEASTER

Phylum - Echinodermata
Subphylum - Eleutherozoa
Class - Echinoidea
Subclass - Irregularia
Order - Clypeastroida
Genus - *Clypeaster*.

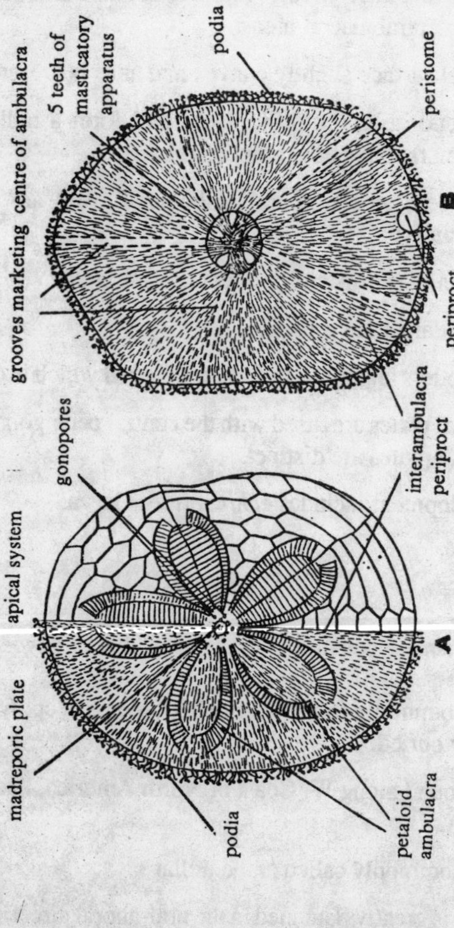

Fig. 1.14. *Clypeaster* A = Aboral view B Oral View

1. It is found on or buries in sandy bottoms of tropical and subtropical seas.
2. It is commonly called as 'cake-urchin'
3. The corona is depressed and disc-shaped, covered all over with short stiff and dark brown spines. Differentiated into oral and aboral surfaces.
4. Oral surface flat and covered with a dense velvet of short and stiff spined and differentiates into five petaloid ambulacral areas and five interambulacral areas.
5. Aboral surface slightly convex and anus is present eccentrally.
6. Madreporite is centrally placed and form a radiate five wide, well-marked petaloid ambulacral areas.
7. Tube feet on the aboral ambulacral areas are flatenned and respiratory.
8. Arms and ambulacral grooves are absent.
9. Pedicillariae are present.
10. Aristotle's lantern with teeth developed which is of simple type.
11. Genital plates are fused with the central pentagonal plate. Radial oscular plates are distinct.
12. Development includes *echinopluteus* larva.

ECHINARACHINUS

The systematic position is the same as that of *Clypeaster*.

1. It is commonly found about 800 fathoms deep, lying on sand or partly buried.
2. It is found along W. Coast of North America, Pacific Coast and U.S.A.
3. It is commonly called 'sand dollar'.
4. Body is greatly flatenned in the oral-aboral direction and covered with soft fur of delicate spines.

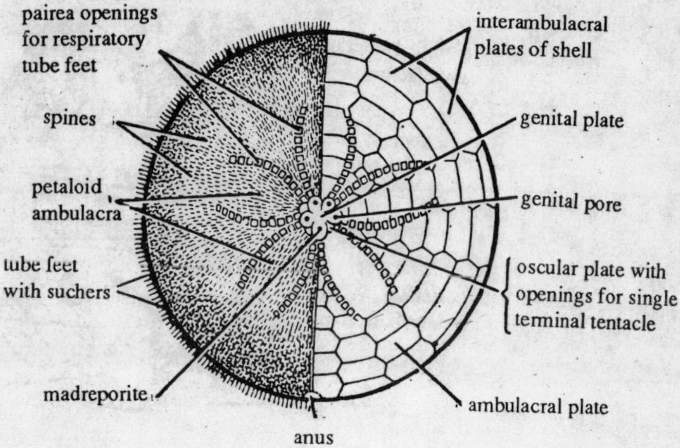

Fig. 1.15 *Echinarchinus*

5. The oral surface is composed of interlocking ambulacral and interambulacral ossicles beset with tiny spines.
6. Mouth centrally situated, surrounded by five petaloid ambulacra. Ambulacra representing opening for tube feet in double row.
7. Madreporite plate, ocular plate and genital plate with genital opening are found on aboral surface.
8. Anus is present on aboral surface.
9. Development includes *echinopluteus* larva.

ECHINOCARDIUM

Phylum	-	Echinodermata
subphylum	-	Eleutherozoa
Class	-	Echinoidea
Subclass	-	Irregularia
Order	-	Spatangoida
Genus	-	*Echinocardium*

1. It burrows in sand. It has world wide distribution.

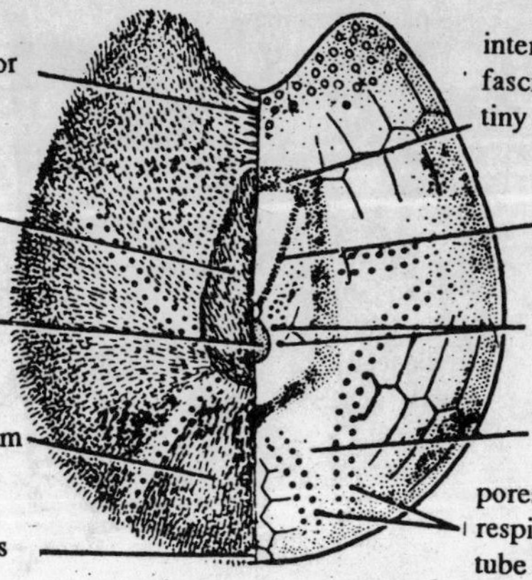

Fig. 1.16. *Echinocardium*

2. It is commonly called 'heart-urchin', as the shape resembles a heart.
3. Body outline cardiform generally light brown in colour.
4. Four petaloids well formed which are flush with test.
5. Mouth eccentric, lies near the anterior end with transversely elongated peristome.
6. Anus situated at the posterior end with periproct.
7. Ambulacra five, petaloid and aborally situated.
8. The Aristotle's lantern is absent.
9. Pedicillariae are of two types with only 2 jaws and scattered among the spines. Poison pedicillariae are absent.

STRONGYLOCENTROTUS

Phylum	-	Echinodermata
Subphylum	-	Eleutherozoa
Class	-	Echinoidea
Subclass	-	Regularia
Order	-	Camarodonta

Genus *Strongylocentrotus*

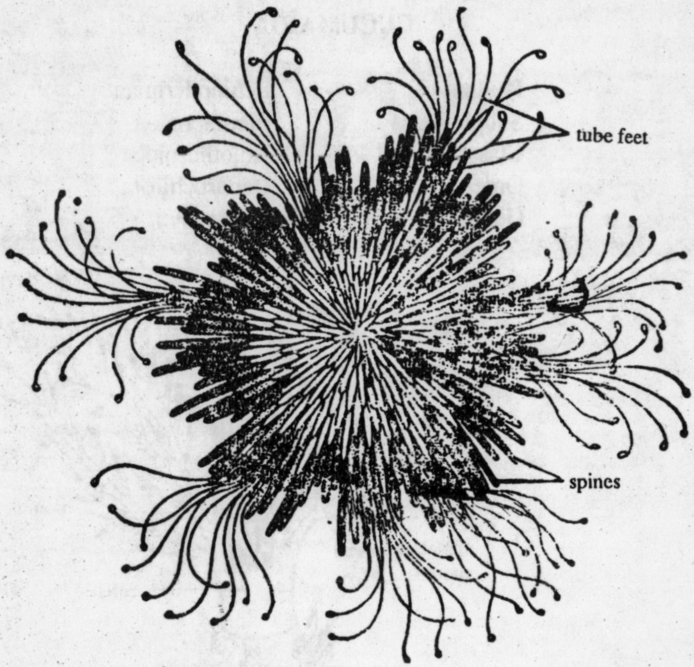

Fig. 1.17. *Strongylocentrotus*.

1. It is commonly called as 'sea urchin;, found in littoral sea.
2. Body is slender in out line and bears slender and short spines.
3. Mouth and anus lie at opposite poles.
4. Body is differentiated into ambulacral and inter-ambulacral areas.
5. Each ambulacral plate bears 7-10 pairs of pores.
6. Tube feet are arranged in five double rows and are greatly extensible.
7. Tubercles are unequal in size and arranged in numerous series and often crowded.

8. Pedicellariae are gemmiform.
9. Gill-clefts are shallow.

CUCUMARIA

Phylum	-	Echinodermata
Subphylum	-	Eleutherozoa
Class	-	Holothuroidea
Order	-	Dendrochirota
Genus	-	*Cucumaria*

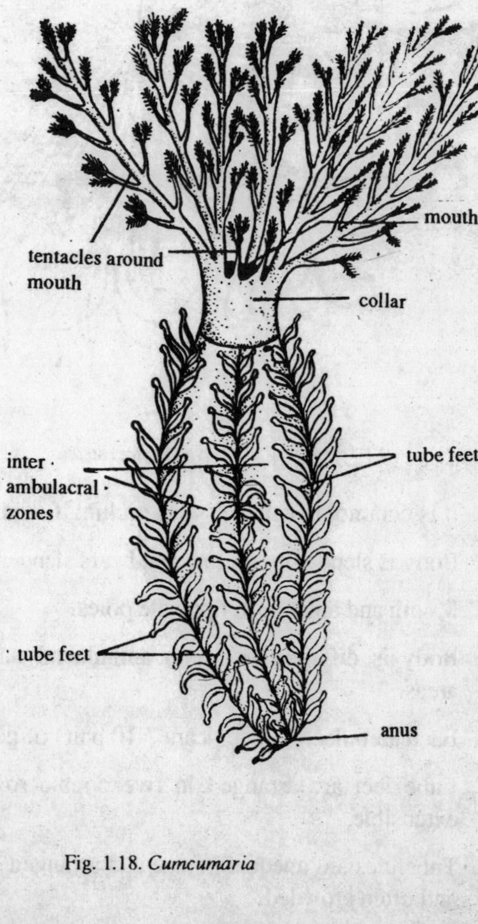

Fig. 1.18. *Cumcumaria*

1. *Cucumaria* is a very common sea cucumber, benethonic and sluggish in habits.
2. Generally feeds on small organisms. Cosmopolitan in distribution and apecially found in Europe, U.S.A. and India.
3. Body is cumcumber-like and is generally cylindrical.
4. Mouth is at broad oral end and the anus at the aboral end which is somewhat narrow.
5. Body is pentamerous, having five longitudinal bands or ambulacra, each provides with locomotory podia.
6. Mouth is surrounded by a lip and peristomial membrane having circular row often dendritic or tree-like tentacles.
7. This collar-like region at the base of tentacles is knwon as introvert.
8. The genital aperture is situated mid-dorsally in between the bases of two adjacent bivial ambualacral tentacles.
9. Body without spines and pedicellariae but highly muscular, leathery and slimy.
10. The exoskeleton in the form calcareous spicules.
11. Sexes are separate.
12. Development with auricularia larval stage.

THYONE

The systematic position is the same as that of *Cucumaria*.

1. It is free swimming holothurian generally found burried in muddy and sandy bottoms of sea.
2. It is found in U.S.A.
3. Podia or tube feet are distributed over the entire body surface.
4. Mouth is at the oral end and surrounded by introvert and ten dendritic tentacles which can be extended to great length.
5. Two tentacles attached to the mid-ventral ambulacral area, are much smaller than others.

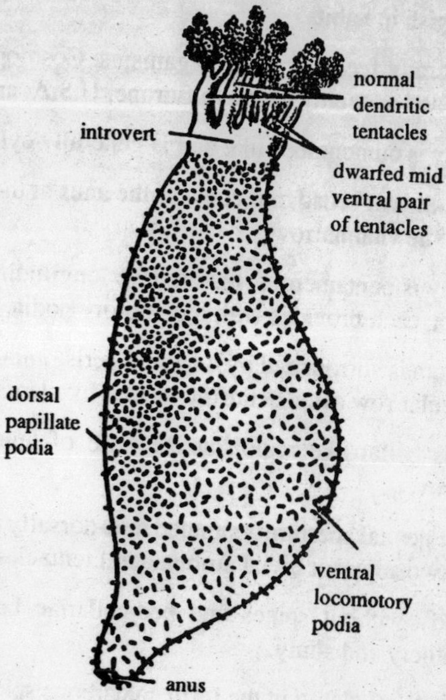

Fig. 1.19. *Thynoe*

6. Anus is present on oboral end.
7. *Thyone* is famous for its great power of regeneration and autotomy.
8. Sexes are separate.
9. Development indirect with *auricularia* larva

ACTINOPYGA

Phylum	-	Echinodermata
Subphylum	-	Eleutherozoa
Class	-	Holothuroidea
Order	-	Aspidochirota
Genus	-	Actinopyga

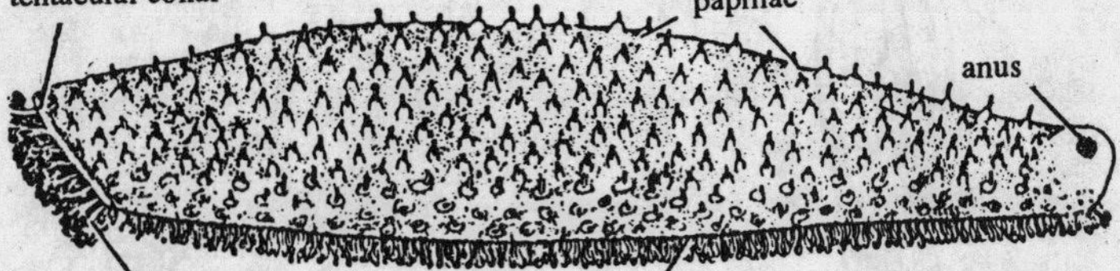

Fig. 1.20. *Actinopyga*

1. It is found in all the sea waters and commonly called sea cucumber.
2. Body is large cylindricle.
3. The entire ventral creeping sole is crowded with numerous locomotory tube-feet.
4. The warts covering the dorsal surface also lack any definite arrangement.
5. The anus is placed at the extreme aboral end. Five calcareous teeth are present within the rim of anus.
6. From the base of left respiratory tree arise a few or numerous pink *Cuvierian tubules* which are emitted from anus, when stimulated.
7. The tubules are toxic to fishes and other animals.

SYNAPTA

Phylum	-	Echinodermata
Subphylum	-	Eleutherozoa
Class	-	Holothuroidea
Order	-	Apoda
Genus	-	*Synapta*

1. *Synapta* is generally found completely burried in soft, sandy soil of the sea bottom.
2. Found in Europe, Pacific coast and Bermuda.
3. Body is elongated, vermiform with rough or warty surface.

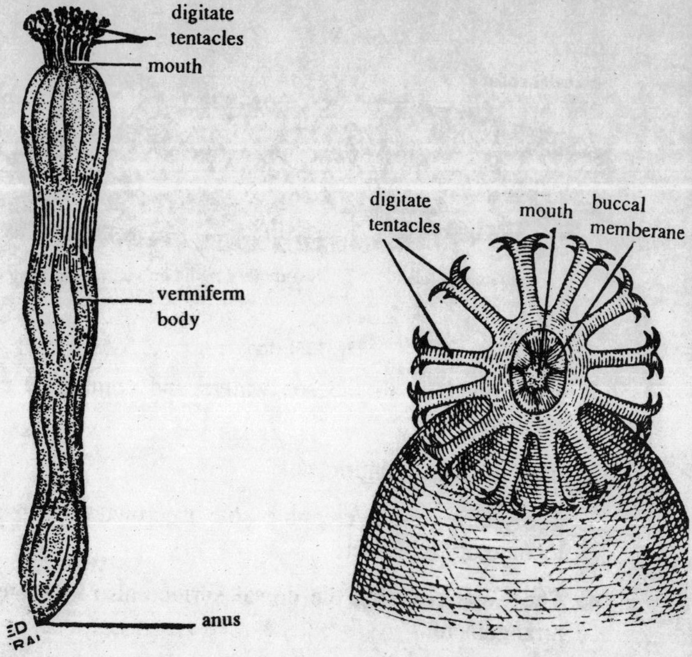

Fig. 1.21. *Synapta*. A = Entire, B = Ora'.

4. Tube feet are absent.
5. Mouth is surrounded by 15 short, digitate tentacles.
6. Anus situated at the aboral end.
7. Respiratory tree and Cuvierian organ absent.
8. Exoskeleton of calcareous spicules are completely absent but with arches and anchor plates.
9. Hermaphrodite. Gonads consist of one pair of branching tubules.
10. Development indirect.

2

ANTEDON

SYSTEMATIC POSITION

Phylum	-	Echinodermata
Subplylum	-	Pelmatozoa
Class	-	Crinoidea
Order	-	Articulata
Genus	-	*Antedon*

The crinoids are the most ancient and in some respects the most primitive of the living classes of echinoderms, flourished during Paleozoic era, and some 80 species still exist today. However, modern sea lilies live at depths of 100 meters or more and are therefore not commonly encountered. The majority of living crinoids belong to a more modern branch of the class, the suborder Comatulida. The comatulids, or feather stars, are nonsessile, free-swimming crinoids that live from the intertidal zone to great depths, and some occur in large numbers on coral reefs. There are approximately 550 species centered primarily in Indo-Pacific waters. *Fishelson* (1974) reports 14 species from reefs in the Gulf of Aquaba distributed in zones down to 50 meters and attaining densities of 40 to 105 individual per square meter.

The *Crinoidea* (Gr. *krinon*, lily+*eidos*, form) are brachiate Echinodermata which remain attached during the whole or part of their lives by the aboral apex of the body; the arms are usually branched and provided with pinnules the tube-feet are tentacle-like, without ampullae, and the water pore is always multiple. The oral surface carries interadially placed anus. They are commonly known as the feather-stars or sea--lilies. The feather-stars or sea-lilies differ from other echinoderms in being fixed permanent or temporarily by jointed stalk. The arms arise from a corresponding number of *thecal plates* or *radials* below which there is a circlet of *basis* often with *infra-basal* plates alternating with them. The modern rosy feather-star (*Comatula or Antedon rosacea)* leave their stalk at a certain stage in life, but the other crinoids, e.g. *Pentacrinus* are permanently stalked like almost all the extinct stone-lilies or encrinites once so abundant. Most of them live in deep water and many in the great abysses. An anchorage is found on rocks and stones, or in the soft-mud, and great number grow together. The free comatulids swim graceful by bending and straightening their arms and they have grappling cirri on the aboral side where the relinquished stalk was attached. By these cirri they move themselves temporarily. Small organisms-diatoms, protozoa, minute crustaceans-are wafted down ciliated grooves on the arms to the central mouth, which is of course on the unturned surface. Some members of the class *(Comatula)* are infested by minute parasitic worms *(Myzostomata)* allied to chaetopods, which form gall on the arms. A lost arm can be replaced, and even the visceral mass be regenerated completely within a few weeks after it has been lost. It has been suggested that occasional expulsion of the visceral sac frees the crinoid from parasites *(Dendy)*.

External Structure

The body of existing crinoids is composed of a basal attachment stalk and a pentamerous body proper, called and *crown*. A well-developed stalk is present in sea lilies but has largely disappeared in the free-swimming feather stars. In the sessile sea lilies, the stalk may reach almost 1 meter in length but is usually much shorter. However, there are fossil species with 25-meter stalks. The basal end bears a flattened disc or, in some cases, rootlike extensions by which the animal is fixed to the substratum.

The stalk (also called *stem* or *column*) is of cylindrical or polygonal contour and is supported by internal skeletal pieces, visible through the surface. The length of the stem in the present stalked crinoids is about 50 cm (about 2 feet). The stalk is variously ornamented and bears many appendages called cirri in many cases, the stalk may be without cirri. In the feather-stars the stalk is lost during development but usually the cirri persist forming one to several circlets springing directly from the aboral surface of the crown.

The cirri show great diversity inform and size. They are short, stout and curved in feather-stars living among arborescent growths, but are long, stout and rigid in those that hold on to rocks, while in those living on muddly bottoms the cirri are slender and straight. The number of cirri varies from 15 to 35. In some cases they may be well over 80. The cirri usually terminate in a claw and may have distally a series of projecting spines.

The pentamerous body (the crown) is equivalent to the body of their lives by the aboral of other echinoderms and, like those of the asteroids and the ophiuroids, is drawn out into arms. The crown is attached to the stalk by its aboral side; thus, in contrast to other living echinoderms, the oral surface is directed upward. The skeletal ossicles are best developed in the aboral body wall, usually called the *calyx*, which is thus somewhat cup like. The oral wall (tegment) forms a more or less membranous covering for the calyx cup, although it may contain embedded calcareous plates. The *mouth* is located in or near the center of the oral surface; five ambulacral grooves extend peripherally from the mouth to the arms. The tegmen is thus divided into five ambulacral and five interambulacral areas. The anus opens onto the oral surface and is usually located in one

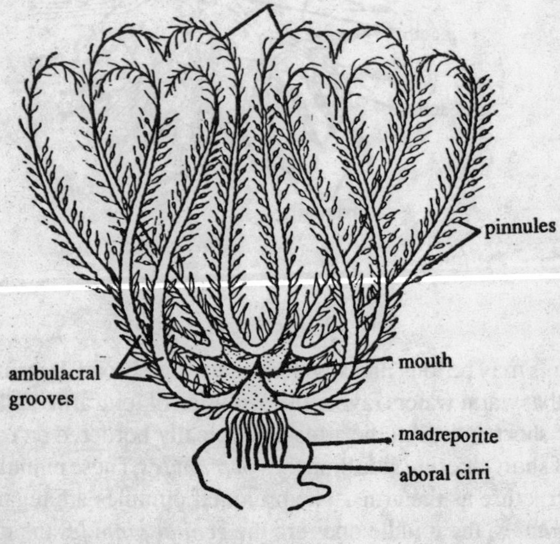

Fig. 2.1. *Antedon*.

of the interambulacral areas at the top of prominence called the *anal cone.*

The arms issue from the periphery of the crown and have a jointed appearance like the stalk. Although there are some primitive species that possess 5 arms, in most crinoids each arm forks immediately upon leaving the crown, forming a total of 10 arms. Further branching results in additional arms. Forty arms is the maximum number in stalked crinoids. However, some comatulids possess 80 to 200 arms. The arms are usually less than 10cm. in length but may reach almost 35 cm. in some species.

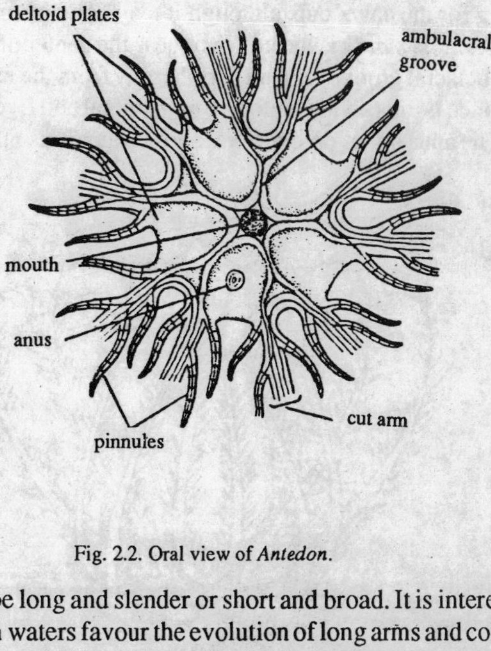

Fig. 2.2. Oral view of *Antedon.*

Arms may be long and slender or short and broad. It is interesting to record that warm waters favour the evolution of long arms and cold water favours short arms. All the arms are typically bordered on each side by a row of short tapering side branches or *pinnules.* These pinnules have the same structure as the arms. The proximal pinnules are regarded as the *oral pinnules,* the middle ones are the *genital pinnules* the rest of them are referred to as *distal pinnules.* The oral pinnules are devoid of ambulacral grooves and podia and act as tactile and protective devices. The genital pinnules contains the gonad and swell during sexual maturity. The distal pinnules are mostly long and slender and become rapidly short near the arm end. Pinnules are generally more or less spiny distally.

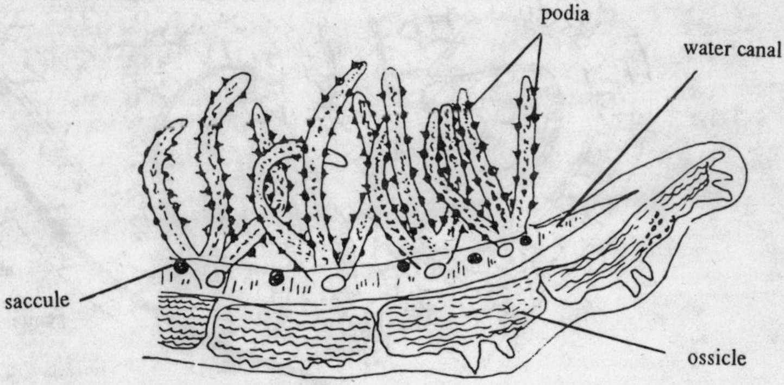

Fig. 2.3. *Antedon*. Pinnules showing podia

The arms arise from a corresponding number of *thecal plates* or *radials* below which there is a circlet of *basis* often with *infra-basal plates* alternating with them. The ambulacral grooves on the arms are deep furrows that converge to the arm bases. Each amublacral groove has raised edges which form regularly repeated *lappets,* alternating on the two sides. The grooves are accompanied by *podia* or *tentacles*. The podia and lappets become reduced along the terminal grooves. The podia along the mouth are termed oral or *labial podia*.

Body Wall

In the region of the ambulacral grooves, the epidermis is composed of ciliated columnar epithelium, but elsewhere the epidermis is for the most part a thin, nonciliated syncytium that is poorly demarcated from the underlying dermis and may be lacking altogether. Most of the dermis is occupied by skeletal ossicles. The stalk, cirri, arm, and pinnules are of a solid construction, being composed almost entirely of a series of thick discshaped ossicles; this accounts for the jointed appearance of these appendages.

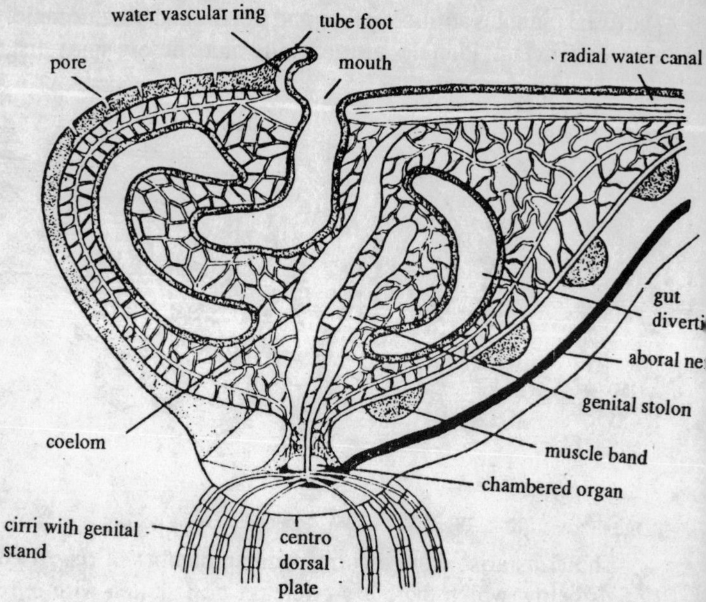

Fig. 2.4. *Antedon*. Selection passing through the central disc

The oral and peripheral surfaces of the ossicles articulate to permit at least some movement, similar to the ossicles composing the ophiuroid arms. The ossicles of the stalk are more securely interlocked than those of the arms, but even here some bending is possible. In more or less the center of the appendicular ossicles, there is a small canal through which runs a nerve and also, in the case of the stalk and cirri, an extension of the coelom. Also, the arm and the pinnule ossicles each bear a deep groove on the oral surface in which is situated the ciliated ambulacral groove.

The ossicles are bound together by distinct connective tissue bands called *ligaments*. The ligaments must possess some contractile powers, since the stalk and cirri are capable of bending movements. Muscles are present in the arms and pinnules and act antagonistically to the ligaments. The position and elasticity of the ligaments is such that they cause the arms to extend. The pair of large muscles that extend between the ossicles and are located toward the oral side bring about flexion of the arms toward the oral side the crown.

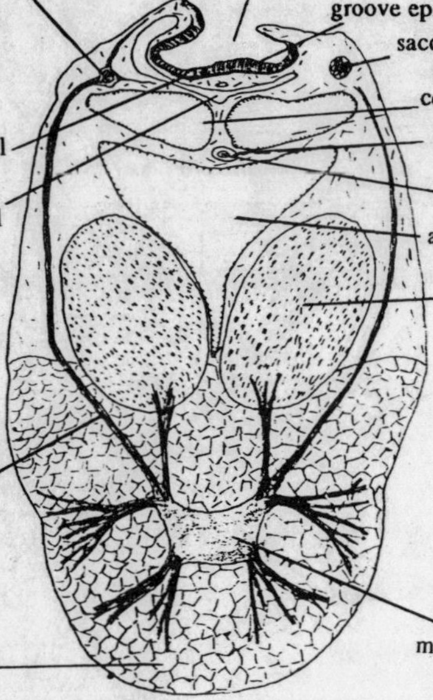

Fig. 2.5. *Antedon*. T.S. through arm.

Locomotion

The sessile sea lilies are limited to bending movements of the stalk and flexion and extension of the arms. The stalkless comatulids, however, are free-moving and are capable of both swimming and crawling. The oral surface is always directed upward, and these animals right themselves if turned over. The cirri are strongly thigmotactic and appear to control the righting reflex. If a comatulid is inverted in such a way that the cirri maintain contract with the substratum, righting does not take place. As soon as the cirri are released, the animal turns over.

Swimming is performed by raising and lowering one set of arms alternately with certain others. In the 10-armed species, every other arm sweeps downward while the alternate set moves upward. In species with more than 10 arms, the arms still move in sets of 5, but sequentially. For example, in a 40-armed comatulid,there are 8 sets of arms acting in sequence. Crawling movements are accomplished by the lifting of the body from the substratum and moving about on the lips of the arms. The arms are often used to grasp and pull the animal over irregular and vertical surfaces.

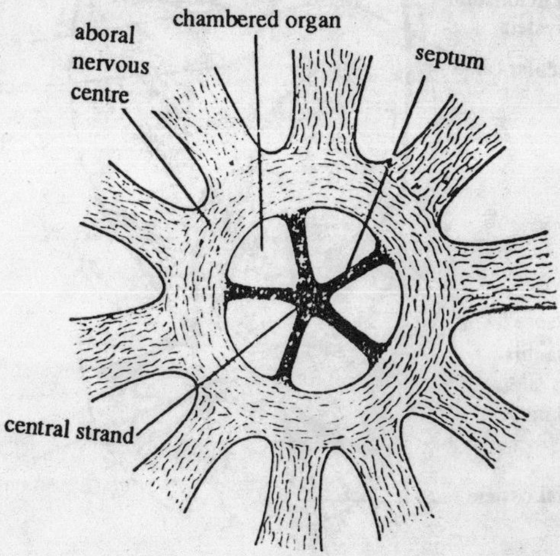

Fig. 2.1. *Antedon*. Digestive System

Feather stars swim and crawl only for short distance and sit perched on the bottom for long periods by means of the grasping cirri. Many shallow-water species are nocturnal. Three Red Sea species of *Lamprometra, Capillaster*, and *Comissia* inhabiting coral reefs hide during the day in crevices and deep within branching corals, keeping their arms tightly rolled. Stimulated by the lowered light intensities at sunset, they crawl upward out of their hiding places to exposed position, where the arms are extended to feed. Species inhabiting deeper water may be very stationary. *Fishelson* (1974) reports 200 specimens of two species of *Decametra* and *Oligomentra* clinging to a gorgaonial coral at about 30 meters for several months in the same position. They roll their arms in response to daytime illumination.

Nutrition

Crinoids are suspension feeders. During feeding, the arms and pinnules are held out stretched and the podia are erect. The podia are shaped like small tentacles and bear long, slender, mucus-secreting papillae along their length. Planktonic organisms or detritus that come in contract with the podia are trapped in mucus and tossed into the ambulacral grooves by the sudden whiplike action of the podia. The cilia of the ambulacral grooves beat in an oral direction, carrying food down the arms, across the disc, and into the mouth. Studies on some Red Sea crinoids disclosed that the food was largely zooplankton in size range of several hundred microns.

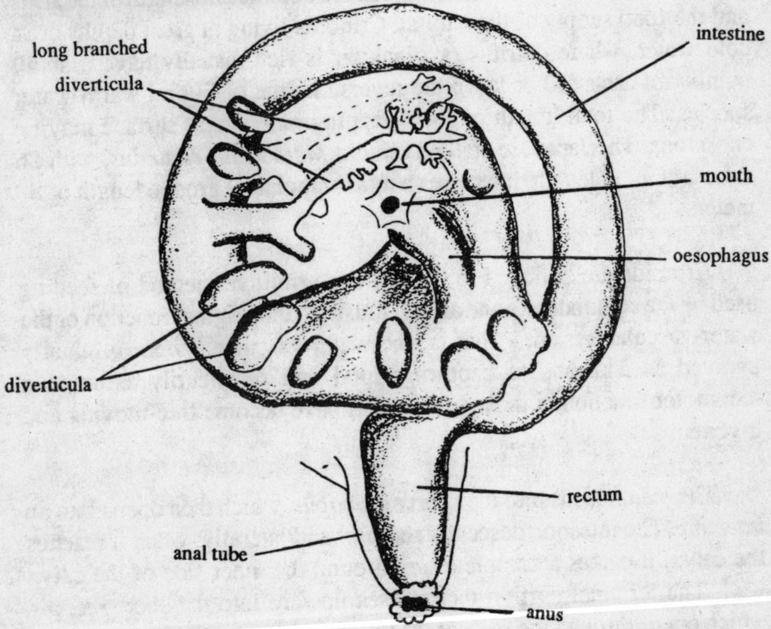

Fig. 2.7. *Antedon*. Digestive System

The feeding position of many crinoids is one in which the arms form a planar vertical filtration fan that is oriented more or less at a right angle to the prevailing current or a circular fan that is tilted into the current. Sea lilies achieve the tilt by bending the stalk. The arm tips may be turned back toward the current so that the animal has the shape of an umbrella. Shallow water comatulids may employ a vertical fan held at a right angle to surf currents, or they may remain in more protective positions within crevices or coral beds and extend the arms in several directions. Some species, such as the European *Antedan bifida,* hold their arms in various positions depending upon the water currents of the habitat in which they live (*La Touche,* 1978)

Only some deep-water crinoids hold the arms in the form of an upward-directed collecting funnel.

There is some correlation between the number and length of the arms and the food supply of the habitat. Crinoids living at great depths or in cold water, where detritus or plankton is rich, usually have a small number of arms (10 or less); the reverse is true of littoral warm-water species. The total length of food-trapping ambulacral surface may be enormous. The Japanese stalked crinoid *Metacrinus rotundus,* with 56 arms 24 cm. in length, possesses a total ambulacral groove length of 80 meters.

Crinoids are believed to display the primitive method of feeding used by the echinoderms and also to illustrate the original function of the water-vascular system - that is, the water-vascular system originally evolved as a means of capturing food and secondarily assumed a locomotor function in those groups that have become free-moving and inverted.

The *mouth* leads into the short *esophagus*, which then opens into an *intestine*. The intestine descends aborally and laterally; when it reaches the calyx, it makes a complete turn around the inner side of the calyx wall. The terminal portion then passes upward into the short *rectum,* which opens through the *anus* at the tip of the an cone. In most crinoids, the intestine bears a number of outpocketings along its inner side.

The details of digestion are still unknown. Wastes are egested as large compact mucusemented balls that fall from the anal cone into the surface of the disc and then drop off the body. The muscular rectum pulsates, but whether this produces a respiratory water circulation through the anus or is an aid to faecal removal is uncertain.

Circulatory System.

The coelom is reduced to a network of communicating spaces as a result in invasion by connective tissue. In the oral side of the arms, the coelom extends as five parallel canals, and in the stalk five coelomic canals pass through the central performation in the ossicles and give off one canal into each cirrus.

The hemal system is a network of spaces and sinuses within the connective tissue strands invading the coelom. There is a definite plexus that surrounds the esophagus, and from this plexus branches extend downward (abroally) through the center of the crown into a spongy mass of cells. This spongy mass is closely associated with the axial gland, which in crinoids is an elongated tubular mesh of glandular tissue occupying the polar axis of the crown. Branches of the hemal system also supply the intestine. There are no hemal space in the stalk, but one hemal sinus extends through the arms, along with the coelomic canals.

Respiration

Gas exchange takes place through any thin part of the body surface exposed to sea water. The podia are undoubtedly an important site, and the great surface area presented by the branching arms makes unnecessary any special respiratory surfaces such as are found in most other echinoderms.

Excretion

Wastes gathered by coelomocytes are believed to be deposited in little saccules located in rows along the sides of the ambulacral grooves. Supposedly, the saccules periodically discharge to the exterior.

Water-vascular system

There is no madreporite in crinoids. The *water ring* encircles the mouth and gives off at each inter-radius a large number of short *stone canals*, which open into the coelom (the feather star *Antedon* possesses about 50 canals at each inter-radius) At each radius of the ring canal, a *radial canal* extends into each arm just beneath the ambulacral groove and forks into all of the branches and into the pinnules. From the radial canals extend *lateral canal* supplying the podia. There are no ampullae, and one lateral canal supplies the cluster of three podia except in the buccal region. Hydraulic pressure for extension of the podia is generated by contraction of radial water canal, which is provided with cross-muscle fibers. Peculiar to crinoids are 500 to 1500 minute ciliated canals (called *ciliated funnels*), which perforate the wall of the tegmen and open into the underlying coelomic spaces. These openings perhaps compensate for the absence of a madreporite, permitting maintenance of proper fluid pressure within the body and therefore indirectly with the water-vascular system.

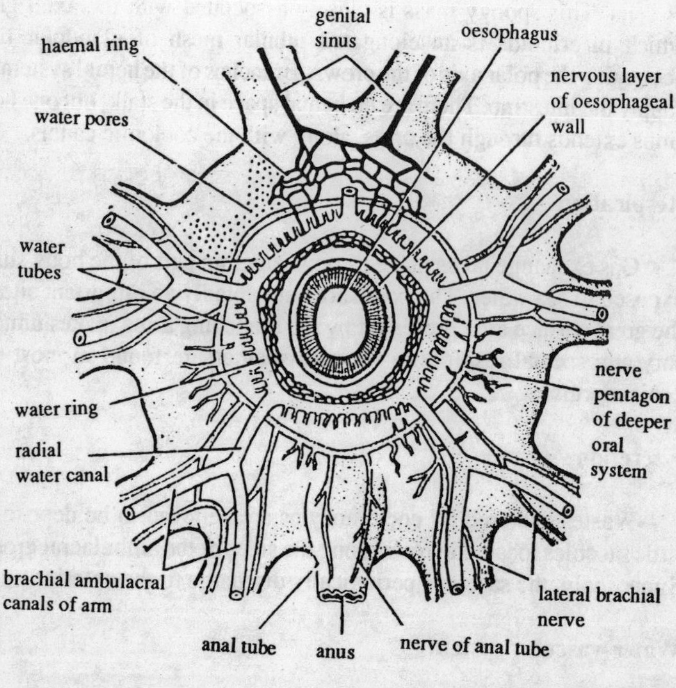

Fig. 2.8. *Antedon*. Water Vascular System.

Nervous system

The crinoids nervous system is composed of three interconnecting divisions. The chief motor system is an *aboral* (or *entoneural*) system located as a cup-shaped mass in the apex of the calyx. From the aboral system, a nervous sheath surrounding the five coelomic canals passes downward through the stalk ossicles and gives off nerves to the cirri along its course,. Also arising from the aboral center are five *brachial nerves*, which proceed through canal in the arm ossicles and supply each

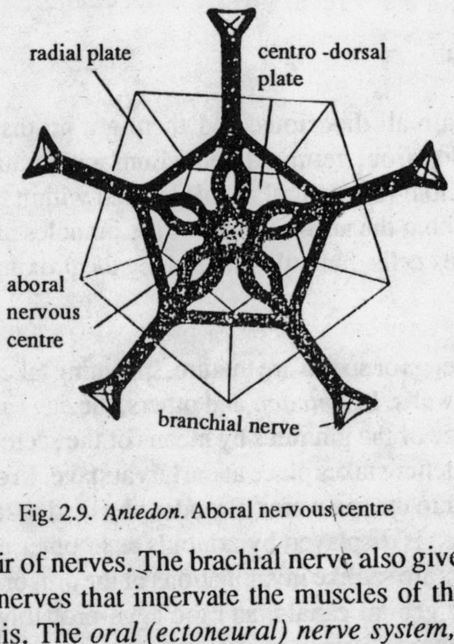

Fig. 2.9. *Antedon*. Aboral nervous centre

pinnule with a pair of nerves. The brachial nerve also gives off along its course a pair of nerves that innervate the muscles of the ossicles and adjacent epidermis. The *oral (ectoneural) nerve system,* which is sensory, is homologous to the principal system of other echinoderms and consists of a subepidermal *radial nerve* that runs just beneath the ambulacral groove in the arms and a nerve ring around the mouth. Just below the oral system is a deeper *hyponeural sensory system* having a *central ring* from which arises a pair of *lateral brachial nerves* to each arm. These nerves innervate the pinnule and podia.

Regeneration

Crinoids possess considerable powers of regeneration and in this respect are requite similar to the asteroids and the ophiuroids. Part of all of an arm can be cast off it seized or if subjected to unfavorable environmental conditions. The lost arms is then regenerated. In same manner, pinnules and cirri are easily regenerated.

In *Antedon*, an animal can replace the loss of one fifth of the disc and the corresponding arms and can sustain the simultaneous loss of four of the five pairs of arms without death. The aboral nerve center is essential in regeneration, and provided this tissue remains intact, all of the visceral mass can be regenerated. The coelomocytes play an active role in regeneration. The coelomocytes play an active role in regeneration. They migrate to the site of the wound in the large numbers, carrying food materials and phagocytizing tissue debris.

Reproduction

Crinoids are all dioecious, and there are no district gonads. The gametes develop from germinal epithelium within an expanded extension of the coelom (the genital canal) located within the pinnules, as in *Antedon*, or within the arms. Not all of the pinnules are involved in the formation of sex cells, but only those along the proximal half of the arm length.

When the eggs or sperm are mature, spawning takes place by rupture of the pinnule walls. In *Antedon* and others, the eggs are cemented onto the outer surface of the pinnules by means of the secretion of epidermal gland cells. Hatching takes place at the larval stage. In other crinoids, the eggs are shed into the sea water. Brooding by cold-water species (many Antarctic forms) is displayed by crinoids as in other echinoderms. The brood chambers are saclike invaginations of the arm or the pinnule walls adjacent to the genital canals, and the eggs probably enter the brood chamber by rupture.

Development

Development through the early gastrula stage is essentially like that in asteroids and holothuroids. As in the latter, the right hydrocoel and the right axocoel never form, although the somatocoel divides into a right and left portion.

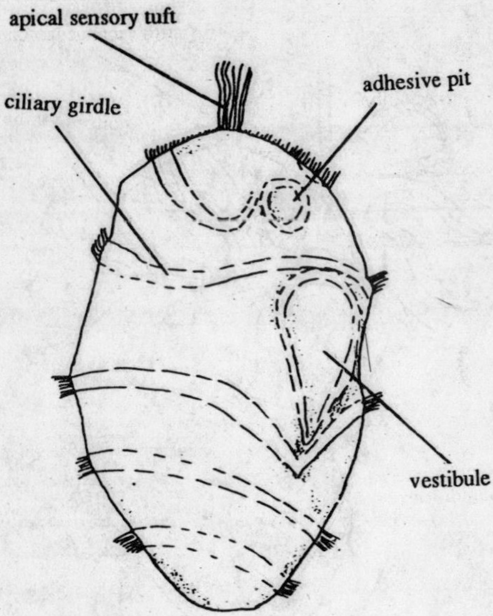

Fig. 2.10. Doliolaria larva.

During the formation of the coelmomisacs, the embryo elongates and development proceeds towards a free-swimming larval stage. The crinoid larva, a nonfeeding doliolaria, is essentially like the vitellaria of holothuroids, being somewhat barrel-shaped with an anterior apical tuft and a number of transverse ciliated bands. In crinoids, the vitellaria is attained directly without the formation of the intervening bipinnaria-like larva of sea cumcumbers. At least in *Antedon,* in which the eggs at spawing adhere to the pinnule surface, hatching takes place at the formation of the doliolaria, which is free-swimming for only a few days at the most.

After a free-swimming existence, the doliolaria settles to the bottom and attaches, employing a glandular midventral depression (the adhesive pit) located near the apical tuft. Then there ensues an extended metamorphosis resulting in the formation of minute stalked sessile crinoids. In the comatulid *Antedon,* actually the only crinoid for which development is well known, metamorphosis also result in a stalked sessile stage (the pentacrinoid larva that resembles a minute sea lily.

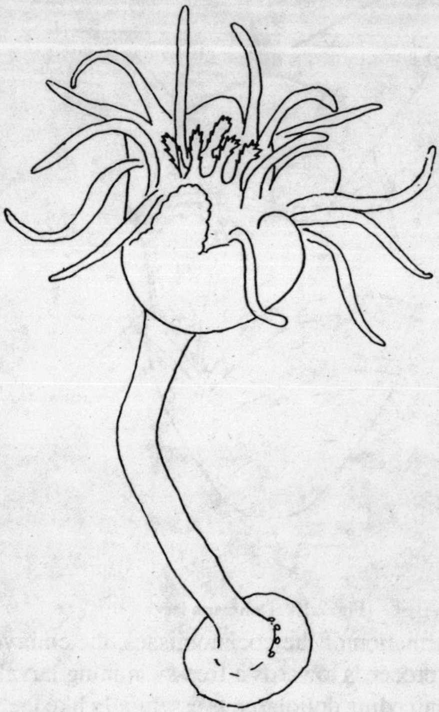

Fig. 2.11. *Pentacrinoid*

The pentactinoid of *Antedon* is little over 3 mm. long when the arms appear, and it requires about six weeks from the time of attachment of the doliolaria to attain this stage. After up to several months as a pentacrinoid, during which time the cirri are formed, the crown breaks free from the stalk, and the young animal assumes the adult free-swimming existence.

3

ASTERIAS

Asterias is commonly called as 'starfish'. The name starfish is somewhat misleading suggesting an organism to be star-like and fish but it lacks in both the characters. However, it is appropriate to call it as sea star as it is star-like in shape and occurs in sea. The gences *Asterias* includes about 150 species.

SYSTEMATIC POSITION

Phylum	-	Echinodermata
Subphylum	-	Eleutherozoa
Class	-	Asteroidea
Order	-	Forcipulata
Genus	-	*Asterias*

Distribution

There are about 1000 species of sea stars and genus *Asterias* includes about 150 species. They have different geographical distribution *Asterias rubens* occurs on the English and North European coasts, *A. vulgaris* on the North Atlantic coast of N.America, *A.forbesi* on the eastern shore from the Maine to the Gulf of Mexico, *A. amurensis* from the Behring sea, Japan, Korea Northward and *A. tenera* occurs on the sea shore from North Scotia to New Jersey.

Habits and habitat

Asterias are free-living and marine echinoderms, crawling about slowly in shallow water to very great depths. It prefers rocky areas where locomotion and concealment are easier. However, the species that live more or less buried necessarily occur on the sandy bottom. The bottom living animals are known as *benthonis*. The most species of *Asterias* are generally solitary but many individuals gather at some place for protection from the direct rays of sun and from excessive drying. Most of them are nocturnal, remain quiet in day time and become active during night. It creeps slowly along the sea bottom. In spite of having a hard integument, it can bend and twist in a variety of ways. They are carnivorous and feed voraciously. They exhibit remarkable power of *autotomy* and *regeneration.*

External Morphology

Shape, Size and Colouration. The body is flatenned in oral-aboral axis and is generally pentamerous in arrangement. The body consists of a central, pentagonal central disc from which radiate out five elongated, tapering, symmetrical spaced *rays or arms.* 6, 7 or 8 rayed specimens are rare and probably result due to regeneration after injury. The body has two distinct surfaces, *Oral* and *aboral.* The oral surface is flat, directed down wards having a centrally placed mouth. The aboral surface is convex, directed upwards and much darker in colour. The oral and aboral surfaces are not the ventral and dorsal surfaces but correspond to the left and right sides of the bilaterally symmetrical larva. The axes occupied by the arms are known as *radii* and the regions of the central disc between the arms are *interradii.* A well defined head is entirely absent.

The size varies from 10-20 cm in diameter. The smallest star fish known is about 1 cm in diameter and the largest 80 inches.

They are usually brightly coloured yellow, brown or orange and purple. The colouration is due to carotenoid pigments.

Oral or actinal Surface

It is the lower flat surface of the body which bears a mouth and usually applied to the substratum. It bears following structures:

1. *Mouth.* On the oral surface, in the centre of the pentagonal central disc is an aperture, the *actinosome or mouth*. It is a

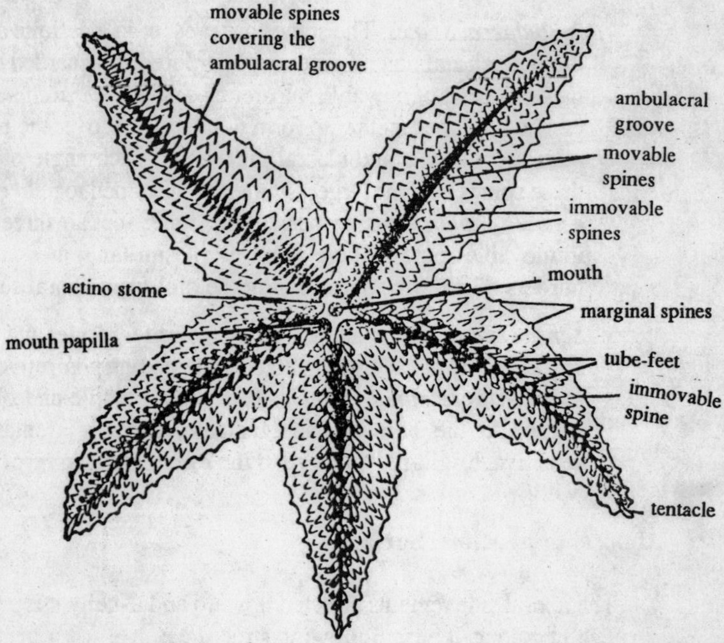

Fig. 3.1. *Asterias*. Oral view.

pentagonal aperture with five angles, each directed towards an arm. The mouth is surrounded by a soft and delicate membrane, the *peristomial membrane* or *peristome* and is guarded by five groups of *oral spines* or *mouth papillae*.

2. *Ambulacral grooves.* Ambulacral grooves radiate out from the five corners of the actionostome and extend one to each arm along its middle upto the distal end. Each groove is bordered by two rows of calcareous *ambulacral spines*. Their bases have muscles, which can bring about the movement of the spines overlapping the tube feet to protect them from injury. External to the ambulacral spine are three rows of large stout immovable spines.

3. *Tube feet.* The tube feet are soft, extensile, tubular processes arranged in 4 radial rows in each ambulacrral groove. Each tube foot is provided with *terminal disc* or *sucker*. The sucker works as suction cup to afford a firm attachment on the surface to which it is applied. These are multipurpose organs as they help in locomotion, food capturing, respiration and as a sensory organs.

4. *Ambulacral Spines.* The spines are present in a definite arrangement. Each ambulacral groove is boardered and guarded laterally by 2 or 3 rows of movable calcareous *adambulacral spines.* They can be brought together to form a latticework over the tube feet retracted into the ambulacral groove. At each angle of mouth, these spines become larger, stouter and grouped together to form a *mouth papilla.* Outside the admbulacral spines are three rows of immovable spines. There is a row of similar spines along the borders of the arms demarcating the oral from aboral surface.

5. *Sense organs.* These include five unpaired *tentacles* and five *eye spots*. The tip of each arm bears a small median non-retractile and sucker-less *terminal tentacle.* It acts as a tectile and olfactory organ. At the base of each tentacle, there is a small, light sensitive, bright-red spot called the *eye*. It is made up of several ocelli.

Aboral or abactinal Surface

The aboral side remains directed upward and usually it is orange or purplish in colour. It bears following structures:

1. *Anus.* It is a small aperture hardly visible with the naked eye. It is situated very nearly, though not exactly, in the centre of the aboral surface, in an interradius next to that occupied by the madreporite.

2. *Madreporite.* The aboral surface is marked by the presence of an asymmetrical subcircular plate known as *madreporite*. It is present in the central disc in one of the interradi close to the base of two arms. The two rays between which the madreporiate is placed are termed the *bivium,* remaining three are called *trivium.* Madreporite introduces bilateral symmetry, because it is only through one vertical plane, which passes through the middle of the madreporite as well as the middle of the opposite arm, that the

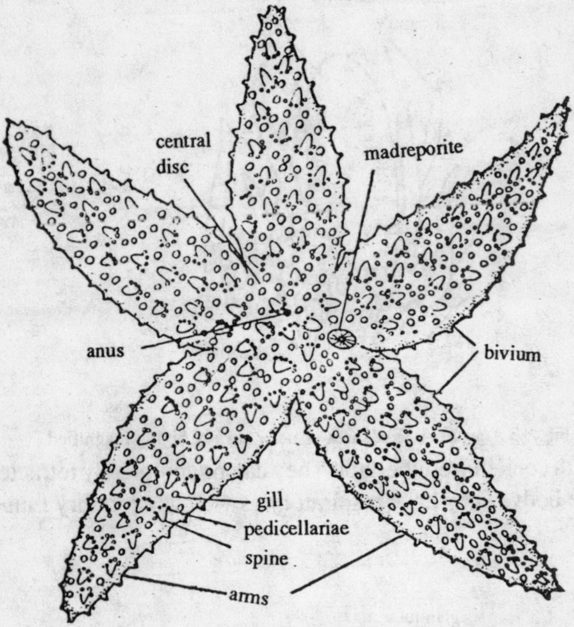

Fig. 3.2. *Asterias*. Aboral view.

Asterias can be divided into two equal, right and left, halves. The surface of madreporite bears a number of radiating, narrow straight or wavy grooves with pores in them.

3. *Spines*. The entire aboral surface is also covered with stout, blunt calcareous spines or tubercles. They arise from the calcareous plate or *ossicles*. The ossicles are embedded in the body wall and covered by epidermis forming the endoskeleton. The spines are arranged in irregular rows parallel with the long axis of the rays.

4. *Dermal branchiae*. The dermal branchiae or gills or papulae are very small, delicate, tubular or conical, finger-like or thread-like, thin walled processes protruding through minute pores in the integument between the ossicles on the aboral surface. The papulae are hollow evaginations of the body wall and their lumen remain in continuation with the coelom. Internally they are lined

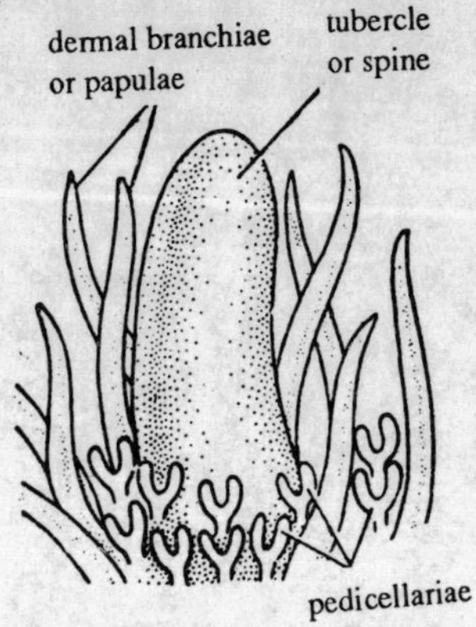

Fig. 3.3 *Asterias* A Small area of body surface highly magnified.

with coelomic epithelium. They can be completely retracted into the body. They have respiratory as well as excretory functions.

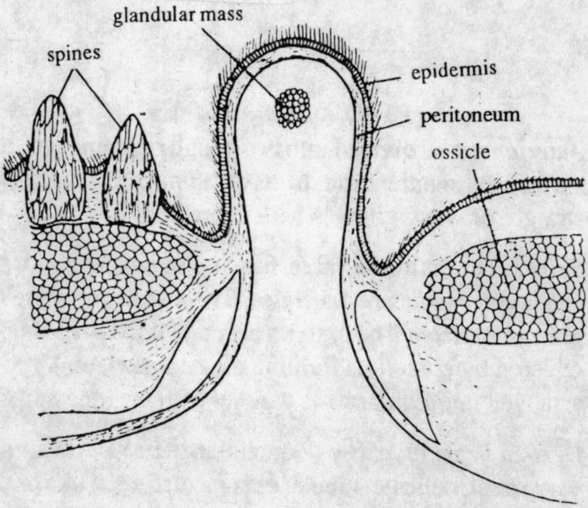

Fig. 3.4. *Asterias* Section through a **Papula.**

Pedicillariae. These structures are present scattered all over the body. These are found in the intervals betweeen the spines of the aboral surface. But, on the oral surface, they are attached to the bases of spines and also in the spaces between them.

These are small whitish, jaw-like or pincer-like structures. They may be stalked or sessile. In *Asterias* only stalked types of pedicillariae are found. Each pedicillaria has a short, flexible and fleshy stalk, lacking internal calcareous support. It posssesses two articulating *calcareous blades* or valves which are supported and articulated by a third calcareous plates is called *basilar or basal piece.* This type of pedicillaria, which is made up of three plates is called *forcipulate.* The opposite surfaces of the blades or valves are serrated and their opening and closing is controlled by one pair of abductor and two pairs of adductor muscles respectively.

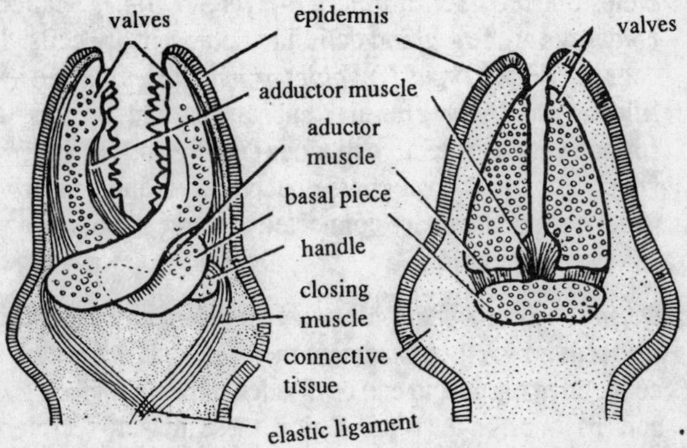

Fig. 3.5. *Asterias.* Pedicillarias Crossed and Straight types.

Asterias Possesses two types of Pedicillariae:

(i) *Straight type or Forceps type.* The two jaws are more or less straight and attached basally to the basal piece. When closed they remain parallel and meet throughout their length like a forceps. These are closed by adductor muscles and open by abductor muscles.

(ii) *Crossed type or Scissors type.* The basal ends of two jaws cross each other as scissors, so that the basal piece is enclosed between their crosssed portions. In this type of pedicillariae, the jaws are also operated by adductor and abductor muscles.

Pedicellariae are sensitive to contact and serve to caputure small prey. They also remove debris and minute organisms such as various larvae sponges and coelenterates which settle on body surface and disturb the respiration by covering the dermal branchiae and tube feet.

Body wall

The body wall of *Asterias* is made up of epidermis, dermis, muscular coat and coelomic epithelium.

1. *Epidermis.* It is continued over all the external appendages. It is composed of ciliated columnar epithelium. The epithelium secretes outside itself a thin two-layered *cuticle.* Scattered in the epidermis are few gland cells and neurosensory cells. The gland cells are of two types (i) *goblet or mucous gland cells* which are filled with finely granular substances and provide the body surface with a protective mucous coating and (ii) *muriform* cells, which are filled with coarse spherules. Neurosensory cells are spindle shaped and are connected to nerve plexus or net situated below the epidermis.

2. *Dermis.* Below the epidermis is dermis which is separated from the former by a thin basement membrane. It is formed of fibrous connective tissue and can be divided into two regions (i) the outer and (ii) inner. In the outer region, these are self-secreted calcareous plates or ossicles which form the hard endoskeleton. It is to be specially noted that in other invertebrates the endoskeleton is ectodermal in origin (if present) where as in *Asterias* it is mesodermal. The inner region of the dermis consists of smooth or involuntary muscle fibres. It is differentiated into an outer circular muscle layer and an inner longitudianl muscle layer. The longitudinal muscle is thicker and better developed on the aboral side than the oral side and brings about bending of the arms.

4. *Coelomic epithelium.* The inner most layer of the body wall is the parietal coelomic epithelium, consisting of ciliated epithelium.

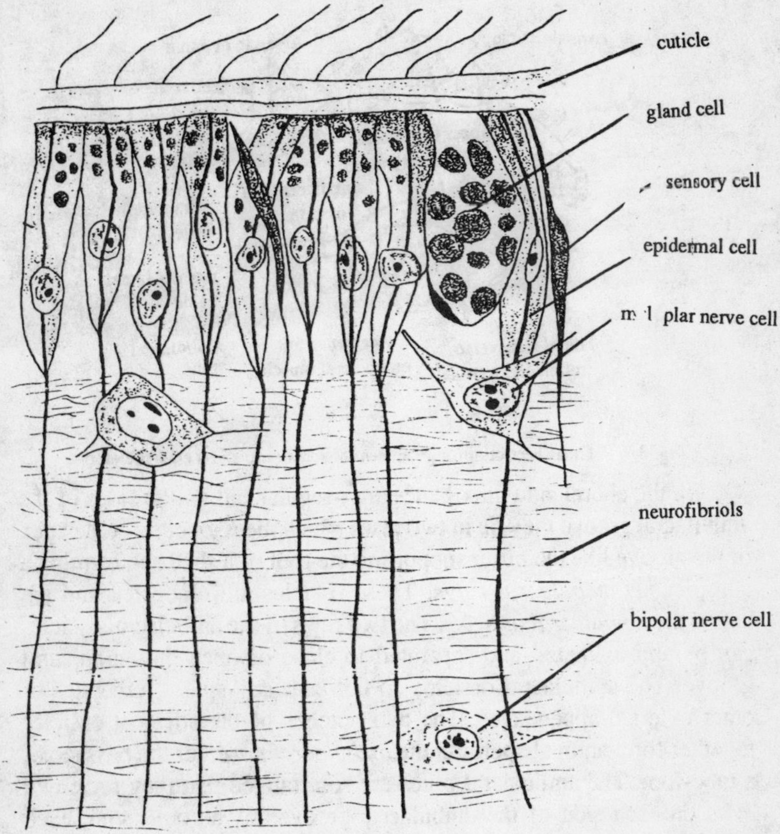

Fig. 3.6. *Asteroas*. Section of body wall

Enoloskeleton

The rigidity of body of *Asterias* is due to the presence of definite skeleton. As already stated, the endoskeleton consists of rod-like calcareous ossicles are of various shapes and are bound together by connective tissue. They form a reticulate skeleton, leaving spaces for the emergence of groups of papulae. The endoskeleton has a crystalline calcite structure. The flexibility of the body results from the fact that the ossicles are not united into an entirely rigid covering, but remain distinct and bound together by connective tissue and muscle that is why movements are possible.

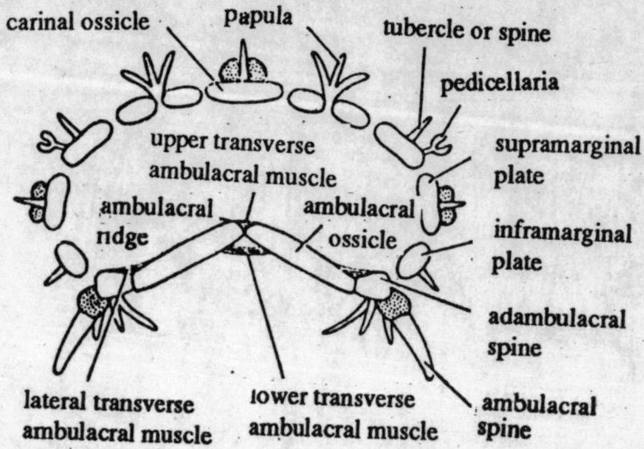

Fig. 3.7. A Cross section through an arm of sea star to show endoskeleton

On the aboral side the ossides are irregular but on the sides of the ambulacral groove they lie in two rows which meet together at the apex of the groove like the rafter supporting the roof of a shed and forming a conspicuous *ambulacral ridge*. These ossicles are elongated and are called the *ambulacral ossicles*. The two rows of the ambulacral ossicles can be approximated and separated to close or open the ambulacral groove. The ambulacral ossicles do not bear any spines, tubercles or other external appendages. The two notches of the adjacent ossicles together form an oval aperture, the *ambulacral pore*, for the passage of a tube-foot. The ambulacral pores are so arranged that they form two rows on each side of the ambulacral groove. At its outer end, each ambulacral groove articulates with one *adambulacral ossicle* forming the edge of the groove and bearing two or three movable spines on small tubercles. Next to the adambulacral ossicle there are two rows of the ossicles called *supra* and *inframarginal ossicles.*

Coelom.

The coelom is present in the form of a large cavity situated below the body wall and extends into the arms. It is lined by ciliated epithelium. These is a large *perivisceral coelom* extending into the central disc and the rays and surrounding the visceral organs. Besides, there are also a number of other coelomic compartments such as the water vascular system, axial sinus, perihaemal canals and sinuses, and the genital

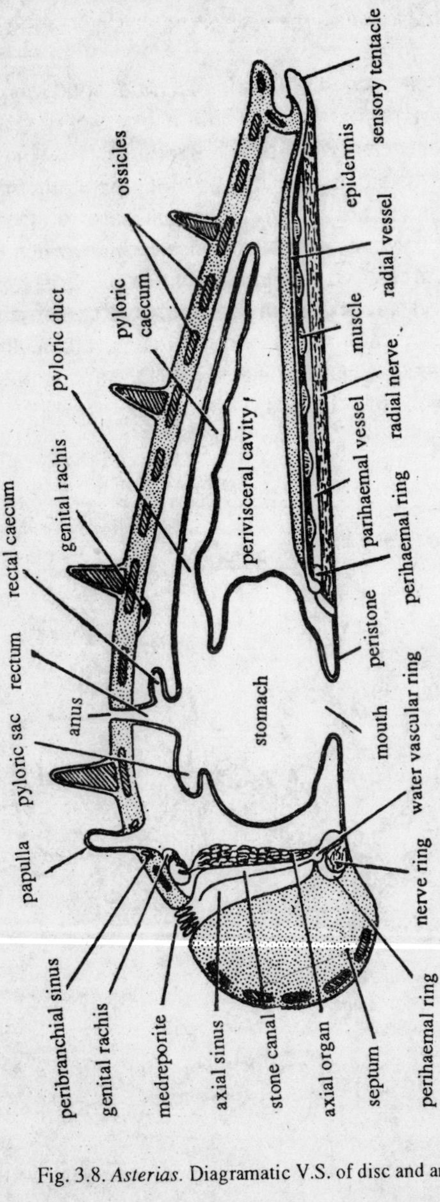

Fig. 3.8. *Asterias*. Diagramatic V.S. of disc and an arm

sinuses etc. These are spaces in the dermis which are without definit lining and, therefore, not regarded of true coelomic origin. The coelomic epithelium is not only applied to the inner surface of the body wall as *parietal layer* but also invest the viscral oragans as *visceral layer*.

The coelom is filled with *coelomic fluid* containing *amoeboid corpuscles*. The composition of this fluid is like that of sea water. The coelomic fluid, like haemolymph of arthropods, serves to bathe the tissue in the absence of true blood. The cilia of coelomic epithelium serve for transportation of materials between various parts of the body. The amoeocytes collect the excretory substances which, when collected in sufficient quantity, is expelled out through some papillae. The amoebocytes also act as phagocytes for removing harmful foreign substances. Some coelomocytes have respiratory pigment. Oxygen is taken into the coelomic fluid and carbon dioxide eliminated into the surrounding sea-water through the thin walls of skin gills.

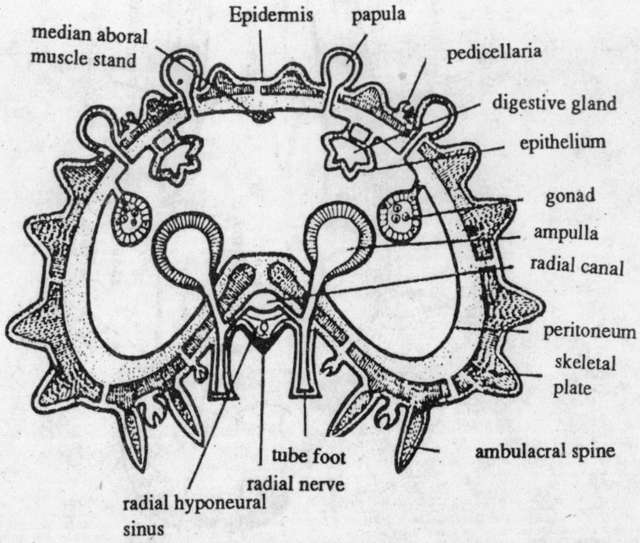

Fig. 3.9. *Asterias*. T.S. of an arm

Digestive System

The digestive system consists of alimentary canal and digestive glands.

I. ***Alimentary canal.*** The alimentary canel is complete extending from oral to abroal side. Therefore, it is very short and wide at places. It comprises mouth, oesophegus, stomach, intestine and anus.

1. *Mouth.* The mouth is the anterior most aperture of alimentary canal and it is situated in the centre of the peristomial membrane on the oral surface. It is provided with a sphincter muscle and radial fibres and is capable of great expansion and retraction. The mouth leads upward into the oesophagus.

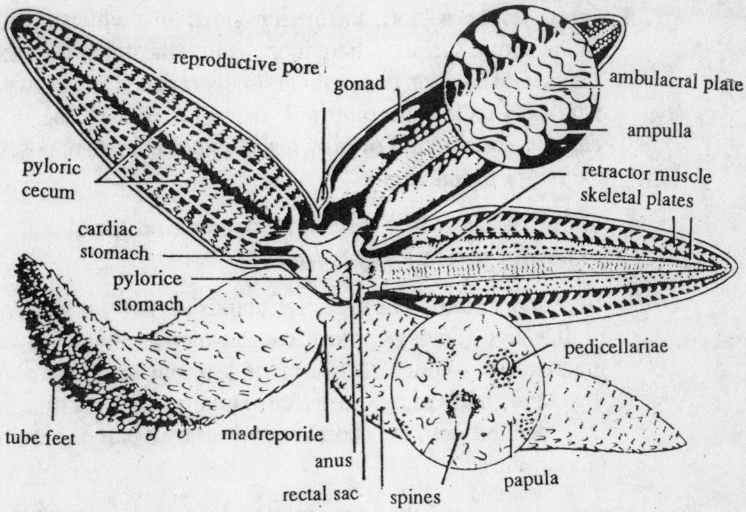

Fig.3.10. *Asterias.* Showing anatomy.

2. *Oesopnagus.* The mouth leads into oesophagus which is a short and wide. It extends ventrally and open into the stomach.

3. *Stomach.* The stomach is a broad sac and fills the interior of the disc. It is typically divided by a horizontal constriction into a voluminous oral part, the *cardiac stomach* and a flatenned aboral part the *pyloric stomach.*

(a) *Cardiac stomach.* It is the larger oval part connected with the oesophagus. Its wall is thin and greatly folded. The cardiac stomach is held in position by five pairs of *mesenteries* or *retractor muscles or the gastric filaments* which are formed of connective tissue and muscles. A pair of *gastric ligaments* connect the cardiac stomach to the ambulacral ridge in each arm. The stomach is capable of everting out through mouth by the action of muscles extending from the body wall.

(b) *Pyloric stomach.* It is a small, flat pentagonal sac opening aborally into the intestine. The angles of the pentagon lie along the radii and each receives a duct, called the pyloric duct, from the corresponding *pyloric caeca* or *digestive glands.*

4. *Intestine.* It is a short, narrow five sided tube which is continued to the aboral surface where it opens into the anus. It gives out at interradial position two or three small branched and brownish appendages called the *intestinal* or *rectal caeca.* The intestinal caeca secrete a kind of brownish fluid which is supposed to be excretory in function.

5. *Anus.* The anus is a very small opening on the aboral surface somewhat eccentrically located.

II. ***Digestive glands.*** The digestive glands of *Asterias* are five pairs of digestive glands or pyloric caeca which are long, brownish or greenish in colour. Each arm of sea star contains a pair of digestive glands or pyloric caeca extending up to its tip, and each is suspended from the aboral body wall by a pair of longitudinal mesenteries.

The hollow axes of the two pyloric caeca of an arm open into a common pyloric duct which opens into the pyloric stomach at one of its angles.

The pyloric caeca are lined by ciliated columnar epithelium having four types of cells:

(i) *Secretary or granular cells* which produce proteolytic, amylolytic and lipolytic enzymes.

(ii) *Mucous cells* which secrete mucus,

(iii) *Storage cells* which store reserve foods as lipids, glycogen and protein-polysaccharide complex.

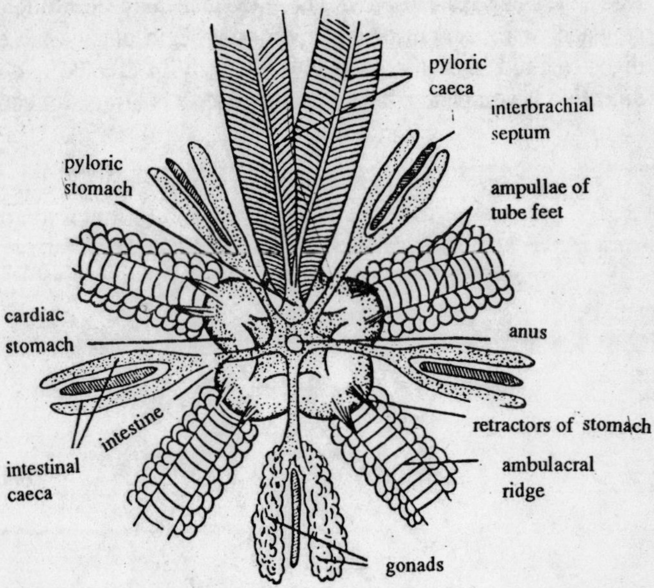

Fig. 3.11. *Asterias*. Showing digestive and reproductive systems.

(iv) *Current producing cells* which bear longer flagella that maintain a steady flow of enzymes towards the stomach along the aboral side and of digested food into the pyloric caeca along the oral side.

Food

The *Asterias* are carnivorous and feed upon all sorts of invertebrates, especially snails, bivalves, crustaceans, polychaetes, other echino derms and even fishes. They also consume the bodies of dead animals encountered on the bottom.

Feeding mechanism, Digestion and Absorption. In some sea-stars, the prey is swallowed whole and digested within the stomach, although the stomach wall must be in contact with the tissues being digested. Shell and other indigestible materials are then eleminated out of the mouth. In *Asterias* through the contraction of the body wall muscles, the coelomic fluid exerts pressure on the cardiac stomach, causing it to be everted through the mouth. The everted stomach, which is anchored by gastric ligaments, engulfs the prey. The prey then may be brought into the stomach by retraction, or digestion may begin outside the body. The digested food in then passed into the body in ciliated gutters. When digestion is completed, the stomach muscles contract, retracting the

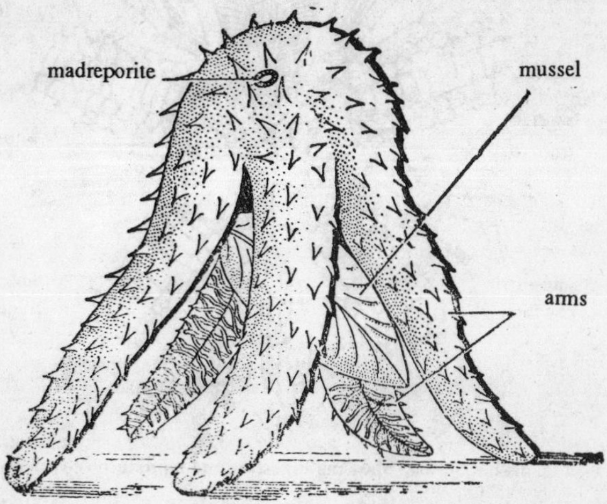

Fig. 3.12. *Asterias*. Feeding on a mussel.

stomach into the interior of the disc. The clams are devoured in an interesting manner. During feeding the sea star extends itself over a clam, holding the gape of the clam upward against the mouth and applying the arms against the sides of the clam valves. The sea star inserts the everted stomach through minute openings between the imperfectly sealed edges of the valves. The gape is produced quite rapidly and not by causing the clam to fatigue over a long period. The everted stomach pours out powerful proteolytic enzymes which completely dissolve the soft viscera of clam. The cilia of stomach set up strong currents that carry the digested food into the stomach and then pyloric caeca. By the contraction of muscles the everted stomach is withdrawn into the interior of the disc.

Thus the digestion is extracellular. Some intracellular digestion is believed to take place in pyloric caeca. The food is chiefly absorted in pyloric caeca, which also store the food material. Distribution of food of various parts of the body takes place through the coelomic fluid. The undigested materials are eleminated mainly through the mouth. Little, if any egestion takes place through the anus.

Water Vascular System or Ambulacral System

The ambulacral system is derived from the coelom, from which it separates during development of the animal. It can be distinguished into following parts:

1. Madreporite
2. Stone canal
3. Ring canal
4. *Tiedmann's bodies*
5. Polian vesicle
6. Radial canal
7. Transverse canal
8. Tube feet

1. *Madreporite.* The *madreporite* is a light coloured, sieve like, circular skeletal plate present inter-radially on the aboral surface of the disc. Its surface is marked by a number of radiating narrow, straight or slightly wavy grooves or furrows which provide it a characteristic appearance. Each furrow is furnished with minute pores which open into a sac-like *ampulla* through minute pore canals. The number of pores may be 250. The ampulla leads into a S-shaped stone canal or madreporic canal.

2. *Stone canal.* The ampulla opens into a S-shaped *stone canal*. The stone canal extends downwards and opens into a ring canal, around the mouth. The walls of stone canal is supported by a series of calcareous rings. The lumen of stone canal is lined by very tall flagellated cells. In embryonic stages and in young *Asterias*, the stone canal remains a simple tube but in adult, lumen

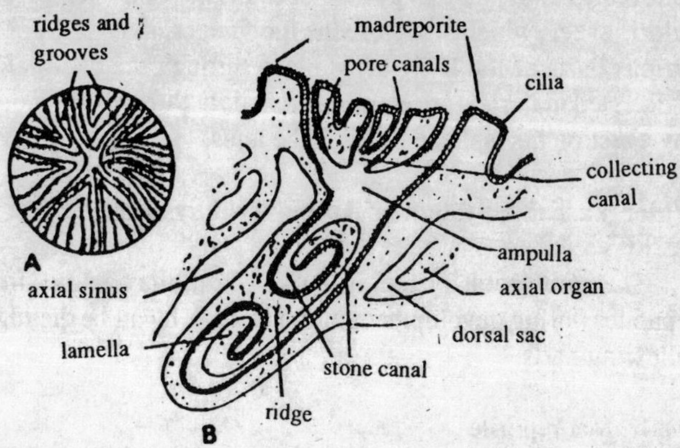

Fig. 3.13. *Asterias*. A=Madreporite seen from outside, b= V.S. through madreporite.

of stone canal possesses a prominent *ridge* with two spirally rolled *lamellae* which by branching become more complicated in structure. During its course, the stone canal along with axial organ is ensheathed by a wide, thin-walled tubular coelomic sac, called *axial sinus*. The three together called *axial complex*.

3. *Ring canal*. It is a wide, somewhat pentagonal canal forming a ring around the oesophagus. The angles of pentagon are situated in the radial position.

4. *Tiedmann's bodies or Racemose glands*. These are small rounded yellow, glandular sacs which open into a ring canal on its inner side, between each radius and interradius except where stone canal opens into the ring canal, there is one Tiedmann's body. Thus there are in all nine Tiedmann's bodies. Each consists of a peritoneum enclosing a stroma of connective tissue and muscle fibres having many radiating tubules. Recently, studies with the electron microscope pointed out that each cell of radiating tubules has flagellum in the midst of microvilli like processes. The function of Tiedmann's bodies is not yet known. However, it is believed that they serve for filteration or as enzyme forming structure or lymphatic glands which produce phagocytic amoebocytes to be set free in the water vascular system.

5. *Polian vesicle*. In most sea-stars, but not in *Asterias*, the ring canal gives off in each interradius one, two or even four large,

thin-walled, pear-shaped sacs known as polian vesicles. They are contractile structures and perform several functions. According to some workers they serve as central organs for regulating pressure in the ambulacral system. According to others they are some sort of lymphatic glands which give rise to amoebocytes of the ambulacral system.

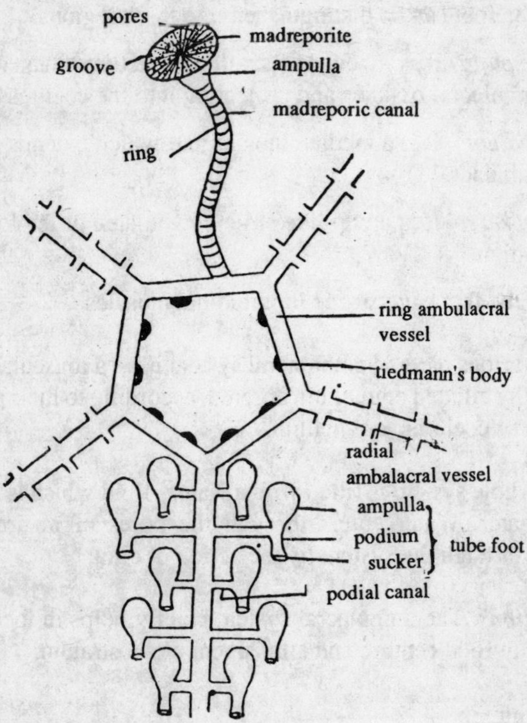

Fig. 3.14. *Asterias*. Ambulacral System.

6. *Radial canals*. From the outer surface of the ring canal are given out 5 radial canals, one to each arm upto its extremity through the ambulacral groove of the arm. Each runs passing above the radial nerve and terminating in the lumen of the terminal tentacle.

7. *Lateral canal*. In each arm, the radial canal gives out two series of short, narrow, transverse branches called *lateral or podial canals*. Each lateral canal is attached to the base of a tube foot and

is provided with a valve to prevent backward flow of fluid the radial canal.

8. *Tube feet.* There are two double rows of tube feet in each arm, one double row in relation to each series of alternately placed long and short lateral canals. Each tube foot has the form of a closed thin-walled tube. It extends through a gap, called the *ambulacral pore,* which lies between two adjacent ambulacral ossiclcs. Each tube foot can be distinguished into three regions.

(i) *Ampulla.* It is a rounded sac-like structure situated above the ambulacral ossicles and projecting into the coelom.

(ii) *Podium.* It is a median tubular part which extends through the ambulacral groove.

(iii) *Sucker.* It is a cup like structure situated at the lower end of podium.

The tube feet have strong longitudinal muscles.

The various parts of ambulacral system have a muscular wall lined internally by ciliated epithelium corered, according to their position, by epidermis or coelomic epithelium.

The whole system is filled with a watery fluid which is identical to the sea water. Water enters through the pores of madreporite and circulates through the system by the action of cilia.

Functions. The ambulacral system chiefly helps in locomotion. It also helps in food cepture and attachment of substratum.

Locomotion

In sea star, there is no head or anterior end. It can move in any direction and any arm can take the lead for locomotion. Locomotion is achieved with the help of water-vascular system which maintains a hydraulic pressure. The beating of cilia of the water vascular system causes the sea water to enter through the madreporite. After passing through the stone canal, ring canal, radial canals and lateral canals, it reaches the tube feet and their ampullae.

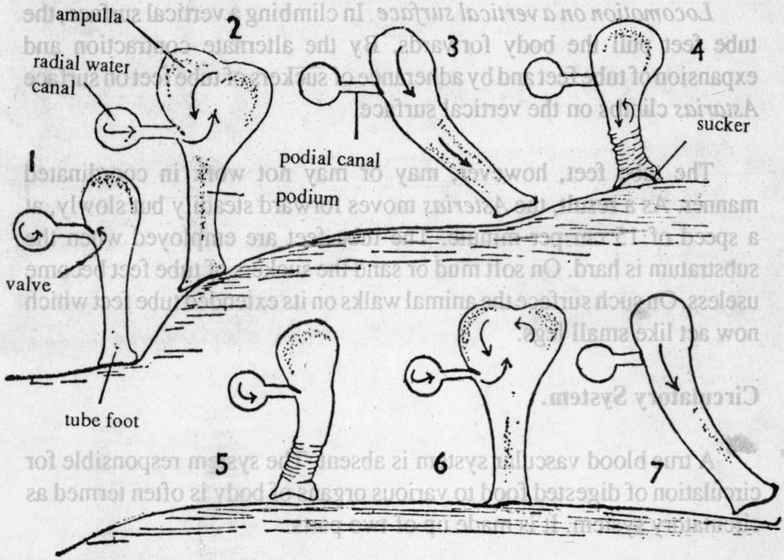

Fig. 3.15. *Locomotion in a star-fish*. 1. Water enters the radial canal, 2. water fills the ampulla, 3. podium swings forwards. 4. podium contracts and vaccum is created below the sucker. 5. sucker is released and podium is pulled back, 6. water fills the ampulla as before, and 7. podium swings forward to catch the substratum.

Locomotion in a horizontal surface. When the sea star desires to more on horizontal surface, the tube feet push the body forwards. The ampullae of raised arms contract, the valve in the podial canal close and the water of ampullae is forced into the podia. The podia elongate in the general direction of movement due to hydrostatic pressure products by the influx of water into them.

The terminal suckers of the tube feet are brought in contact with the substratum and their central parts are withdrawn to form suction cups. The adhession is further strengthened by mucous secreted by the tips of tube feet. The tube feet now pivot forward on their attached suckers, assuming vertical position and thereby pushing the body forward. The tubefeet then shorten by contracting their longitudinal muscles and forcing some water back into their ampullae and releaces their suckers. The ampullae then contract again and whole sequence of events is repeated.

Locomotion on a vertical surface. In climbing a vertical surface, the tube feet pull the body forwards. By the alternate contraction and expansion of tube feet and by adherance of suckers of tube feet on surface *Astarias* climbs on the vertical surface.

The tube feet, however, may or may not work in coordinated manner. As a result, the *Asterias* moves forward steadily but slowly, at a speed of 15 cm per minute. The tube feet are employed when the substratum is hard. On soft mud or sand the suckers of tube feet become useless. On such surface the animal walks on its extended tube feet which now act like small legs.

Circulatory System.

A true blood vascular system is absent. The system responsible for circulation of digested food to various organs of body is often termed as circulatory system. It is made up of two parts.

I Perihaemal System

II Haemal System

I. *The perihaemal system*. It is also derived from the coelom and is lined with cuboidal ciliated epithelium. It comprises:

1. Axial sinus
2. Aboral ring sinus
3. Genital sinuses
4. Oral ring sinus
5. Radial perihaemal sinuses
6. Marginal sinuses
7. Peribranchial sinuses

1. *Axial sinus*. It is a wide, nearly vertical, thin-walled tube surrounding the stone canal and the axial gland. At its aboral end it communicates with the ampulla of stone canal and also with the exterior by some of the pore canals of the madreporite. At the same end, it opens into the aboral ring sinus. The three (stone canal, axial gland and axial sinus) togetheer form an *axial complex*.

2. *Aboral ring sinus.* It is a pentagonal channel lying around the intestine, just beneath the central disc. It does not make a full circuit as it is incomplete in the region of the stone canal. It gives off 10 *genital branches*, two in each inter-radius. The aboral ring sinus and its genital branches contain *genital rachins*, a continuation of the axial gland.

3. *Genital sinuses.* Each gonad is surrounded by a small genital sinus which is connected with the aboral ring sinus by a short and slender brnach.

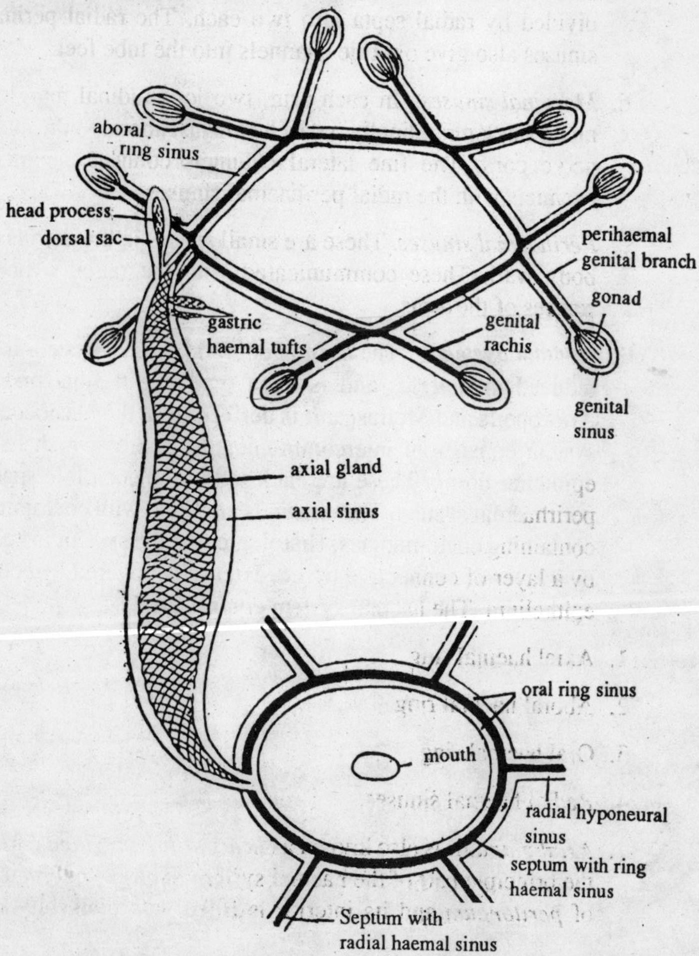

Fig. 3.16. *Asterias.* Perihaemal and haemal System.

4. *Oral ring sinus.* At its oral end, the axial sinus opens into the inner division of a circular channel, the *oral ring sinus* or *peribuccal* or *hyponeural perihaemal ring sinus* surrounding the mouth. It is a large tubular sinus and internally divided by an oblique septum (the haemal strand) into two-a smallar inner ring and a larger outer ring. From the outer part of the oral ring sinus arises five radial perihaemal sinuses.

5. *Radial perihaemal sinuses.* 5 such sinuses arise from oral ring sinus. Each one extends through each arm between the radial nerve and the radial ambulacral vessel. The radial sinuses are also divided by radial septa into two each. The radial perihaemal sinuses also give out fine channels into the tube feet.

6. *Marginal sinuses.* In each arm, two longitudinal marginal sinuses run longitudinally on each side just aboral to the marginal nerve cord. The fine lateral channels connect the marginal channels with the radial perihaemal sinuses.

7. *Perihaemal sinuses.* These are small spaces in the dermis of the body wall. These communicated with the radial hyponeural sinuses of the arms.

II. *Haemal System*. The haemal or blood lacunar system is much reduced in *Asterias* and is open type like the haemocoel of Arthropoda and Mollusca. It is derived from the blastocoel. The system consists of intercommunicating spaces which have no epithelial lining. These are enclosed in the coelomic spaces of perirhaemal system. The channels are filled with coelomic fluid containing coelomocytes. Histologically, the system is bounded by a layer of connective tissue, externally covered by coelomic epithelium. The haemal system comprises:

1. Axial haemal ring
2. Aboral haemal ring
3. Oral haemal ring
4. Radial haemal sinuses.

1. *Axial gland.* It is also known as *heart or brown gland* and forms the principle part of the haemal system. It has an external lining of *peritoneum* and its interior is filled with connective tissue

containing many small inter-communicating spaces filled with a fluid with amoeboid coelomocytes with a brown pigment. The axial gland is connected with the oral haemal sinus at its oral end and with aboral haemal sinus at its aboral end. A small terminal *head process* arises from the aboral end of the axial gland. It communicates with a small dorsal sac in the madreporite. A pair of gastric tufts arising from the haemal sinuses in the wall of cardiac stomach open into the axial gland near its aboral end. The food digested in the stomach enters the haemal circulation through the gastric tufts.

2. *Aboral haemal ring.* The axial gland, at its aboral end connects with the ampulla of madreporic canal and joins a pentagonal ring canal situated beneath the aboral surface of the central disc. From this canal, five pairs of genital haemal strands are extended to gonads. These haemal strands are two in each inter-radius.

3. *Oral haemal ring.* It is the circular haemal sinus, located around the mouth just below the ring canal of the water vascular system.

4. *Radial haemal sinuses.* From the oral haemal ring arise five radial haemal sinuses or strands. One of these runs out through the septum of each hyponeural radial sinus. The radial haemal branches give out from their sides short branches to tube feet.

Respiration

In *Asterias*, respiration takes place by the dermal branchiae or papulae scattered over the aboral surface and by the tube feet projecting in the ambulacral groove. Both the respiratory structures are thin walled and are continuous with coelom. Oxygen dissolved in sea water diffuses into coelomic fluid through these structures. The coelomic fluid distributes oxygen to all internal tissue and brings carbon dioxide to the respiratory organs for outward diffusion.

Excretory System

These is no distinct excretory system in echinoderms. It is likely that there may be some organ in connection with the ambulacral system or with the axial sinus for excretion. The nitrogenous metabolic excretory waste usually contains ammonium compounds. The excretion is brought about with the agency of coelomocytes that pick up wastes and pass them out through the papullae. The rectal caecae serve in excretion to some extent. There is cetrain amount of diffusion of dissolved wates through the gills and thin wall of tube-feet.

Nervous System

The nervous system of *Asterias* is of simple and primitive type. It is confined to the body wall. It consists of four parts:

1. Superficial or Ectoneural or Epidermal nervous system
2. Deeper hyponeural
3. Aboral or coelomic nervous system
4. Visceral nervous system

1. *Superficial or Ectodermal nervous system.* This nervour system is situated just beneath the epidermis. It is the main part of sensory nervous system. It includes circumoral nerve ring, radial nerve cords and subepidermal nerve plexus.

(i) *Circum oral nerve ring.* The nerve ring is pentagonal in shape and is circum-oral, i.e., occurs around the mouth in the peristomial membrane. It supplies nerve fibres to the peristomial membrane and the oesophagus and at each radius gives off a radial nerve.

(ii) *Radial nerve cords.* The radial nerve cords arise from the angles of the nerve pentagon and extend outwards, one through each arm. Each radial nerve terminates as a sensory cushion on the aboral side of the terminal tentacle. Each nerve cord is 'V' shaped in cross-section and continuous on its outer side with the epidermis. It remains separated on its inner side from the hyponeural sinus only by a thin dermis and the coelomic epithelium.

(iii) *Subepidermal nerve-plexus.* It is in the form of an extensive network of nerve cells and nerve fibres situated below the epidermis all over the body surface including the papulae, pedicillariae and spines. In each nerve cord it is joined to radial nerve cord by five nerve fibres. In each nerve cord it forms a pair of marginal nerves out of which one traverses along each margin of the ambulacral groove and forms a nerve ring in the sucker of each tube feet.

The nerve pentagon and radial nerves contain many neurons, arranged in sensory, association and motor tracts. They receive and coordinate responses similar to that of central nervous system found in other animals.

2. *Deeper nervous system.* It is situated on the oral side just beneath the perihaemal sinuses and a little above the epidermal nervous system by a thin layer of connective tissue. It consists of a *double nerve pentagon* running round the mouth and sending a *double radial nerves or Lange's nerve* in each arm. Each nerve is continued as a plate of nervous tissue in the outer oral wall of the radial hyponeural sinus and gives out branches to the muscles of the arm.

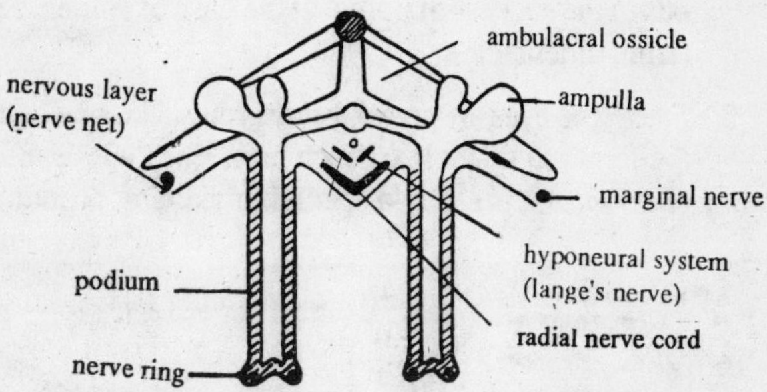

Fig. 3.17. *Asterias*. Nervous System in section.

The deep nervous system originates from the mesoderm. Its function is primarily *motor*.

3. *Aboral or coelomic nervour system.* It is represented by a thin nerve plexus situated in the aboral body wall, just above the parietal layer of coelomic epithelium. It is somewhat thickened to form an *anal nerve ring* in the central disc and a nerve in each arm. The aboral nervous system is connected with the marginal nerves by several *lateral nerves* in each arm. This system is also mesodermal in origin and motor in nature.

4. *Visceral Narvous System.* It is situated in the gut wall just outside the enteric epithelium. It has the form of a conspicuous plexus which has connections with the visceral recepetors and also with the muscles layers of gut wall.

Sense Organs

The sense organs of *Asterias* are of three types :

1 Neurosensory cells

2. Eyes.

3. Terminal tentacles.

1. *Neurosensory cells*. These are present here and there in epidermis. They are more or less spindle shaped containing the nucleus, a distal thread-like process reaching to the cuticle, and a proximal fibre entering the subepidermal nerve plexus. They may act as tectile receptor or chaemoreceptors. They are especially numerous in the suckers of podia, at the base of spines, pedicillaria and terminal tentacles.

2. *Eyes*. The eyes are small, bright-red spots, one under the base of the terminal tentacle of each arm. Each eye is also called *optic cushion* as it is really a special thickening of radial nerve. Each

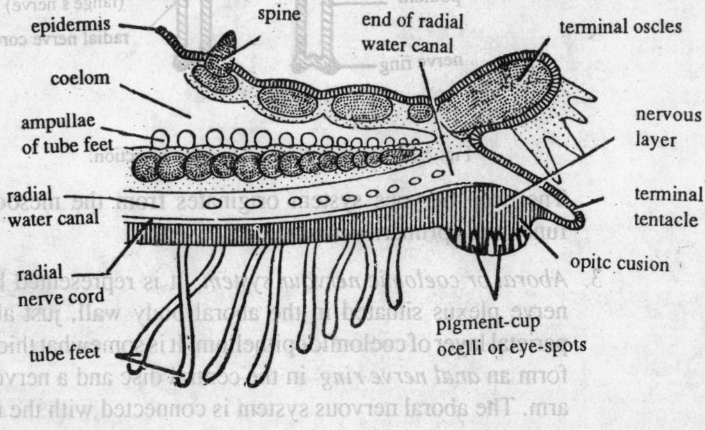

Fig. 3.18. *Asterias*. L.S. of an arm through terminal tentacle showing position of eye.

eye comprises a number of cup-shaped concavitics lined by alternating pigmented cells and retinal cells. The pigment cell

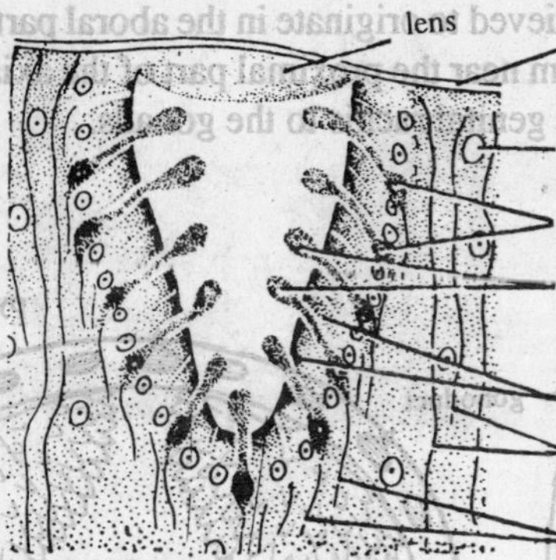

Fig. 3.19. *Asterias*. V.S. through an eye pit.

contain an orange pigment. The retinal cells are elongated cells with a distal bulbous enlargement projecting into the cavity of the cup and a proximal fibre passing into the underlying radial nerve. The number of ocelli in one eye ranges from 80-200 in different species. Each concavity is covered by a transparent cuticle and may contain a lens beneath the cuticle and may contain a lens beneath the cuticle. The lens is formed by epidermis. Eyes are sensitive to light and detect change in the intensity of light.

3. *Terminal tentacles.* The terminal tentacles have sensory cells which are tactile and also sensitive to food.

Reproductive System

Asterias is unisexual i.e. *dioecious*. The male and female are separate but both the types of gonads are morphologically alike and situated in similar positions. They have only microscopic differences. The reproductive system is simple. Ovaries and testes consist of small rounded follicles like the bunches of grapes. They, however, differ in colour. The testes are pale-grey while ovaries from pink to orange. These are five pairs of gonads in either sex, a pair in each interradius. The gonads are attached at the bases of the arms and extend freely into the proximal parts of the arms on the sides of pyloric caeca. From the proximal end of each gonad arises a short dilated *gonoduct* which opens out laterally on the aboral surface by a minute *gonopore*. There are no special organs for copulation. The accessory glands and recepticles are also absent.

The sex-cells are believed to originate in the aboral part of the axial or from the wall of coelom near the proximal part of the axial gland and then migrate through the genital rachis to the gonads.

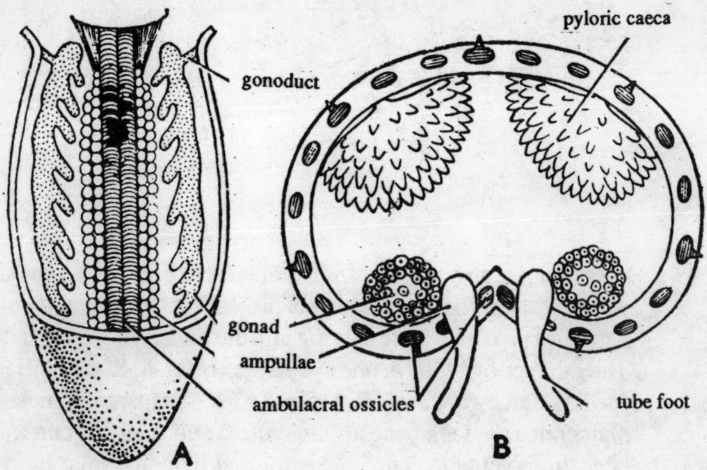

Fig. 3.20. *Asterias*. Gonads.

There is a single breeding season in sea-stars i.e. late spring. The ripe ova and sperms are simply shed into the sea water where fertilization takes place. Maturation and sheding of gametes are controlled by a secretion of neuro-secretory cells in radial canal (*Chaet* and *McConnangly*, 1959). According to *Binyon* (1964), the axial organ may serve as a chaemoreceptor for substances received through the madreporite, thus stimulating the release of gametes. A single female lays about 200 million eggs in a season.

Fertilization. It is external and takes place in the sea water.

Development

After fertilization the zygote divides into two equal blastomeses, the cleavage is *holoblastic*. Cleavage leads to the formation of *blastula* by second day. The blastula sooner or later develops a blastocoel and is called *coeloblastula*. After a free swimming existence of about a day the blastula undergoes embolic invagination and becomes a two layered, cup-like *gastrula*. Its outer layer is formed of *ectoderm* and inner layer

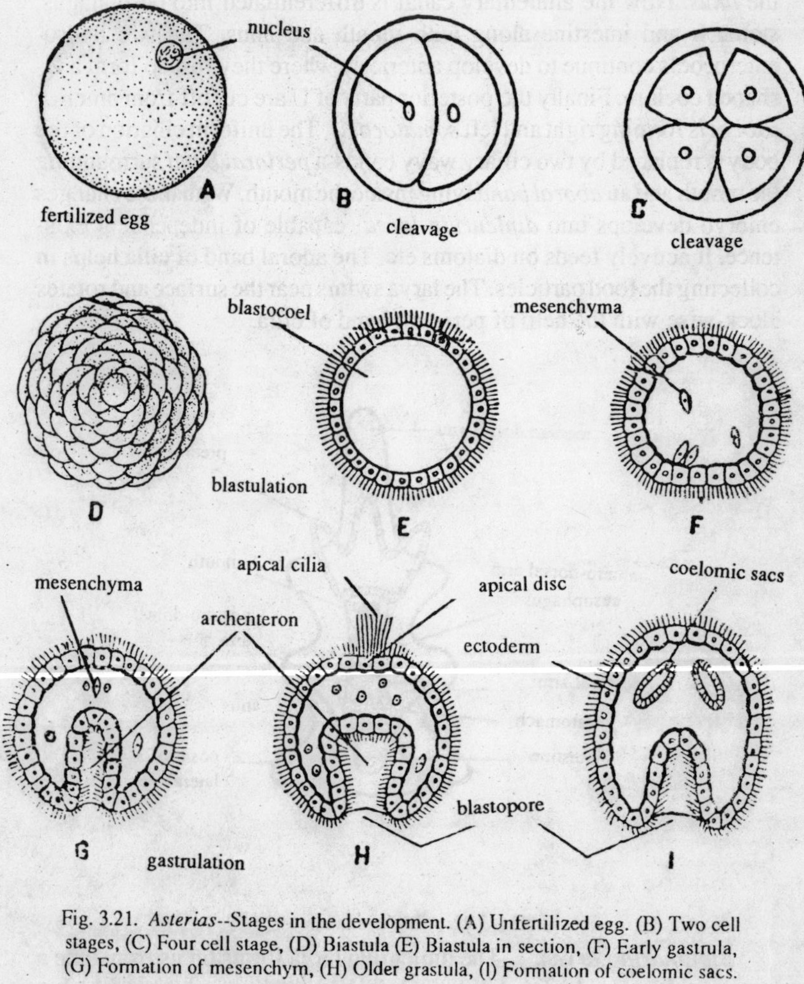

Fig. 3.21. *Asterias*--Stages in the development. (A) Unfertilized egg. (B) Two cell stages, (C) Four cell stage, (D) Biastula (E) Biastula in section, (F) Early gastrula, (G) Formation of mesenchym, (H) Older grastula, (I) Formation of coelomic sacs.

of *endoderm*. The cavity of gastrula which is lined by endoderm is known as *archenteron*. It opens to the exterior by a wide aperture called *blastopore*. The archenteron formed as a result of gastrulation is narrow. When fully formed it gives rise to a coelomic sac, the *hydroenterocoel*, on each side. The two occupy the right and left of the archenteron and gradually elongate anteriorly in the plan of the long axis of the larva. From the left hydro-enterocoel arises a dorsal evagination which forms the water pore. Ventrally a stomodaeal inpushing is formed. This meets the archenteron and completes the digestive tract. The blastopore forms the *anus*. Now the alimentary canal is differentiated into oesophagus, stomach and intestine along with mouth and anus. The two hydro-enterocoels continue to develop anteriorly where they fuse to form a U-shaped coelom. Finally the posterior parts of U are cut off from anterior *axocoels* forming right and left *somatocoels*. The uniform ciliation of the body is replaced by two ciliary wavy bands-a *perioral band* surrounding the mouth and an *aboral band* lying inside the mouth. With these changes embryo develops into *dipleurula larva* capable of independent existence. It actively feeds on diatoms etc. The adoral band of cilia helps in collecting the food particles. The larva swims near the surface and rotates clock-wise with the help of peri-oral band of cilia.

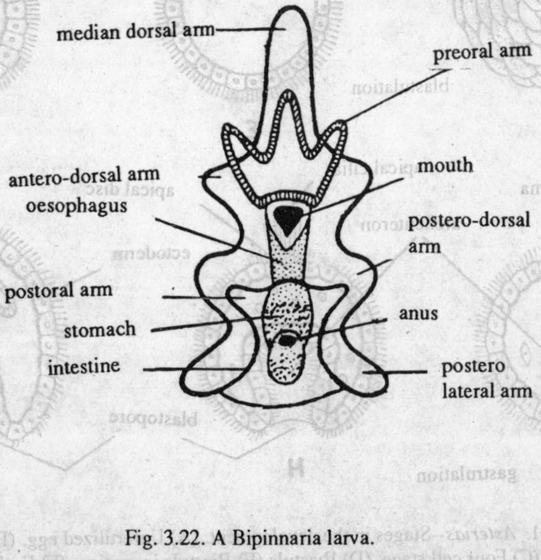

Fig. 3.22. A Bipinnaria larva.

Bipinnaria Larva. The dipleurula larva forms on its front side a big preoral lobe, which is bordered by a *preoral loop* made up of cilia.

Soon three lateral lobes are formed and these lobes are also bordered by a *postoral loop* of cilia. The larva is now called bipinnaria. It is also bilaterally symmetrical. It swims and feeds independently. After some weeks it is transformed into the next larval known as brachiolaria larva.

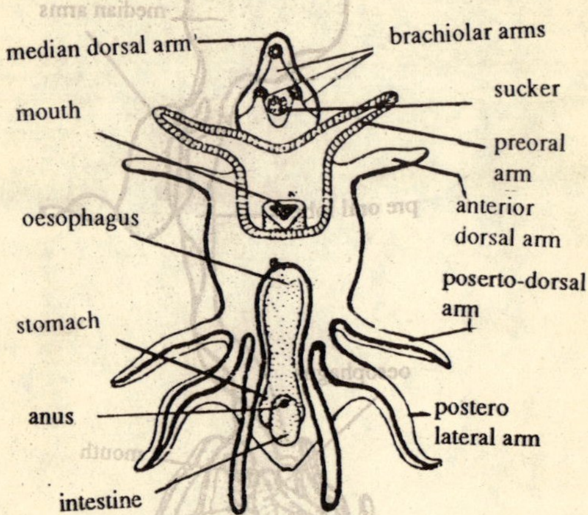

Fig. 3.23. Brachiolaria larva

Brachiolaria larva. The lobes of bipinnaria larva become modified into long, slender, ciliated and contractile *larval arms.* From the preoral lobe arise three short and non-ciliated processes ending into *suckers.* These processes are called *fixing processes.* This is now known as *branchiolaria larva.* It swims and feeds like bipinnaria larva.

Metamorphosis

The bilaterally symmetrical larva gradually metamorphoses into the radially symmetrical adult. After about 6 or 7 weeks the brachiolaria settles down on some solid object or to the bottom, where it remains temporarily attached by its fixing processes. The anterior end of the larva degenerates and forms only an attachment stalk, and the adult body develops from the rounded posterior end of the larva. The left side becomes the oral surface and the right side becomes aboral. The adult arms appear as extensions of the body. Internally the entire digestive tract

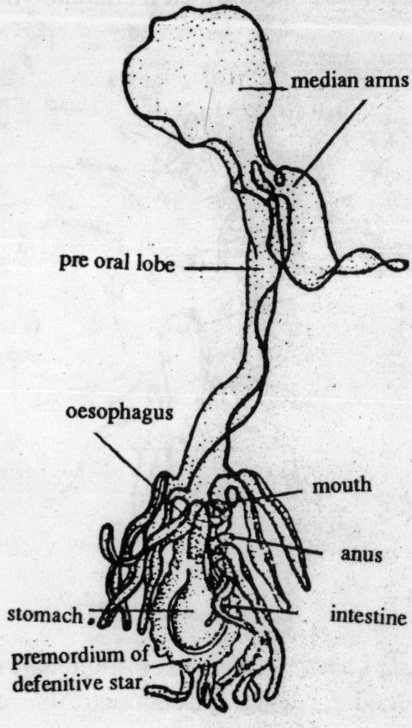

Fig. 3.24. Metamorphosis in larva

degenerate. All these parts formed anew and in a postion coinciding with the adult radial symmetry. The left axohyronceol forms the water-vascular system and from the hydrocoel develop five pairs of projections, two in each of the developing arms. These projections represent the cavity and coelomic lining of first pair of podia in each arms. Asabout this time, the skeletal system appears. The first adult ossicles to be formed are those around the pole aboral pole, which contains the anus. New ossicles are then added peripherally to the initial skeleton. The detached young sea star is not more then 1 mm in diameter.

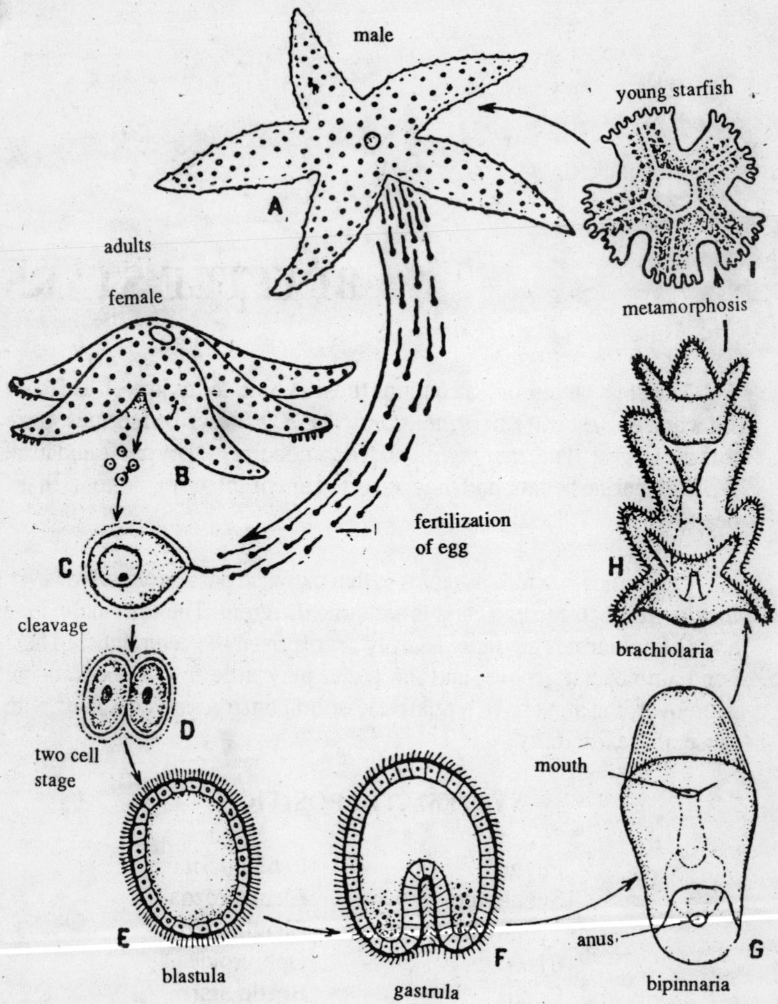

Fig. 3.25. *Asterias* Life History

4

BRITTLE STARS

The class Ophiuroidea contains those echinoderms knwon as *basket stars* and *serpent stars,* or *brittle stars*. The 2000 described species make this the largest of the major groups of echinoderms. They are found in all types of marine haitats, and they are often abunfant of soft bottoms in the deep sea.

Ophiuroids resemble asteroids in that they also possess arms. However, in other respects the two classes are quite different. The extremely long arms of ophiuroids are more sharply set off from the central disc. There is no ambulacral groove, and the podia play little role in locomotion. Moreover, the arms have a relatively solid construction compares with those of the asteroids.

SYSTEMATIC POSITION

Phylum	-	Echinodermata
Subphylum	-	Eleutherozoa
Subphylum	-	Eleutherozoa
Class	-	Ophiuroidea.
Genus	-	Brittle Star

External Structure

Ophiuroids are relatively small echinoderms. The disc in most species ranges from 1 to 3 cm. in diameter, although the arms may be quite long. The basket stars are the largest members of the class and the disc in some species of this group may attain a diameter of almost 12 cm.

A great variety of colors are found in the ophiuroids, but the coloration is often not conspicuous because of their small size and the mottled and banded patterns.

The central disc is falttened, displaying a rounded, pentagonal, or somewhat star-shaped circumference. The aboral surface varies from smooth to granular and may bear small calcareous plates, called *shields*, and small *tubercles* or *spines*.

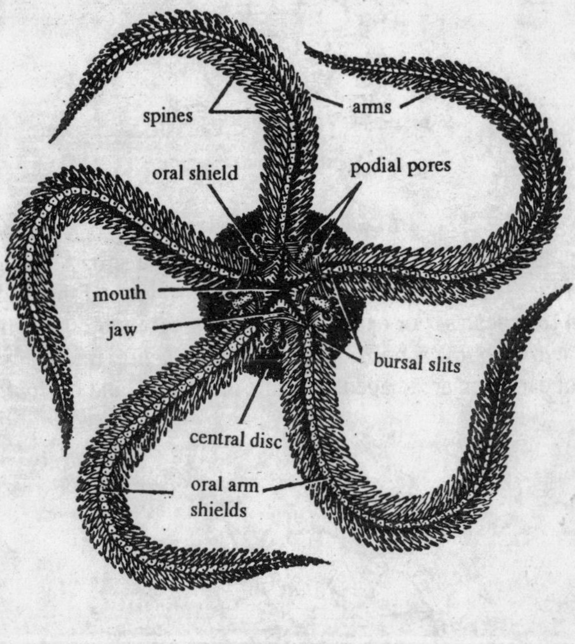

Fig. 4.1. *Ophiothrix*. Oral view.

There are typically only five arms, however, in basket stars the arms branch at either the base or more distally, and the subdivisions repeatedly branch to produice a great mass of coils that resemble tentacles. The arms characteristically appear jointed, because of the presence of four longitudinal rows of shields. There are two rows of lateral shields, one row of aboral shields, and one row of oral shields. A single set - that is, two lateral, one aboral and one oral shield - completely surround the arm and corresponmd in position to an internal skeletal ossicle, which is described later. Not infrequently, the oral and aboral shields are reduced by the large size of the lateral shields, which may even meet on the oral and aboral surfaces.

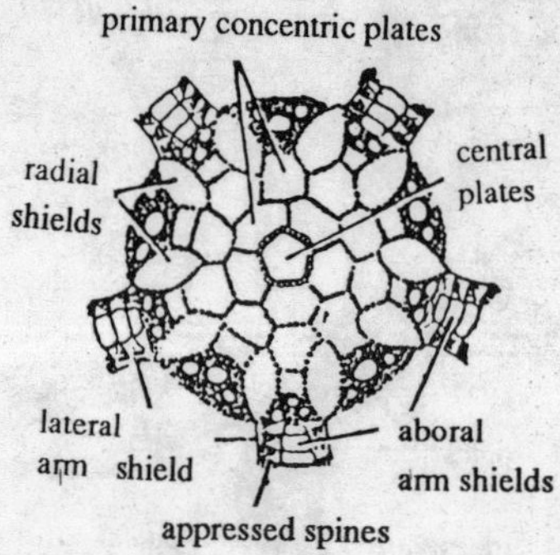

Fig. 4.2. *Ophiolepis*. Disc in aboral view

Each lateral shield usually bears 2 to 15 large spines arranges in a vertical row. These spines vary considerably in size and shape, depending upon the species. For exmaple, in *Ophiothrix* and others, the most oral spine in each row is hook-shaped, in *Ophiohelusa* the spines toward the end of each arm are shaped like little umbrellas, and in *Ophiopteron*

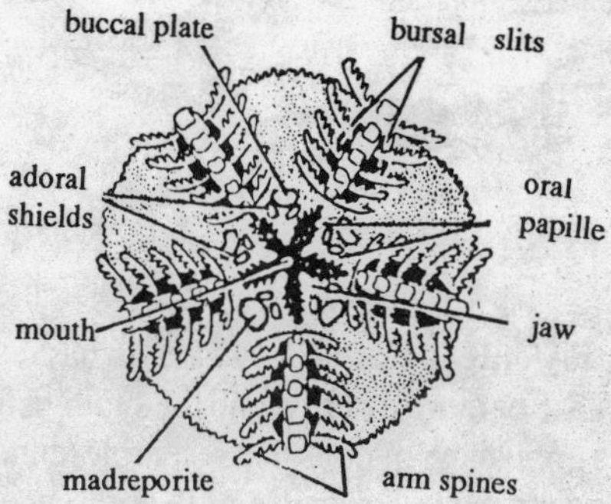

Fig. 4.3. *Ophiura* Disc in oral view.

some of the spines are interconnected by webbing. Some serpent stars have what are probably poisonous spines, for the end of such a spine is provided with a thick glandular covering.

There is no ambulacral groove on the oral surface of the arms. The ambulacral ossicles have sunk inward and enlarges to form the vertebrae, and the radial water canal lies within the ossicles. The ambulacrum is said to be closed. The podia are small, tentacle-like papillate appendages that extend between the oral and the lateral shields, and there is typically one pair of podia per joint. Neither papulae nor pedicellariae are present in ophiuroids.

The center of the oral surface of the disc is occupied by a complex series of large plates that frame the mouth area and also form a chewing apparatus. The plates are organized into five triangular interradial jaws that are oriented longitudinally with the apex directed toward the center of the disc. Each jaw is composed of several plates, of which the most conspicuous are a pair of half jaws and a large basal oral shield. The jaws bear many small toothlike plates along their margin. At the apex, the teeth also extend aborally in a vertical row.

In most ophiuroids, one oral shield is modified, forming a *madreporite*. Thus, the madreporite is located on the oral surface, in contrast to its aboral position in echinoids and asteroids. The arms extend inward to the jaws on the oral surface of the disc, leaving five large, somewhat triangular, interradial areas having essentially the same surface structure as the aboral side of the disc.

Body Wall And Skeleton

With the exception of the basket stars, the epidermis of the ophiuroids is a reduced *syncytium*, continuous with the underlying dermis. Moreover, aside from certain specialized areas, there are no surface cilia. The dermis contains the more superficial skeletal shields, as well as the large deeper ossicles of the arms. Each ossicle is a large disc-shaped skeletal piece (a vertebra) that almost fills the entire interior of the arm.

The vertebral ossicles are arranges linearly from one end of the arm to the other, and each ossicle is covered by four superficial arm shields. The two end surfaces of a vertebral ossicle are structurally adapted for articulation with adjacent vertebrae. In most serpent stars this articulation allows great lateral mobility of the arm but no vertical movement. However, in the basket stars, which have a somewthat different vertebral articulation, the arms can bend and coil in any direction.

In basket stars, there is muscle layer beneath the dermis, but in serpent stars the muscle layer is absent and the coelomic peritoneum covers the inner surface of the dermis. There are, however, well-developed intervertebral muscles for the movement of the arms.

Several ophiuroids, including the cosmopolitan *Amphipholis squamata,* are luminescent. A yellow or yellow-green light emanates from the sides of the arms, particularly in the region of the spines. The light is intracellular within neuron-like photocytes.

Locomotion

The ophiruoids are the most highly mobile of all echinoderms. During movement the disc is held above the substratum, with one or two arms extended forward and one or two arms trailing behind. The remaining two lateral arms perform a rapid rowing movement against the

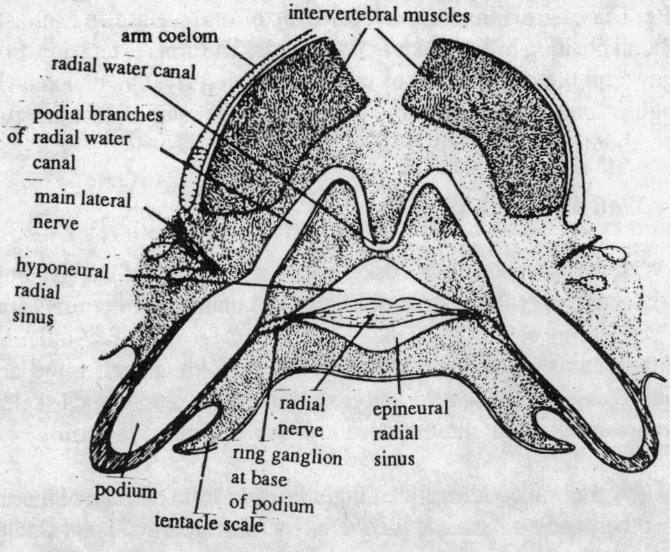

Fig. 4. .4. Section of an ophiuroid arm to show podial water canals.

substratum that proples the animal forward in leaps or jerks. The spines provide traction. Serpent stars show no arm preference and can move in any direction. Righting is accomplished by extending two adjacent arms away from each other so that they form a straight line tangent to the disc. Using the extended arms as a pivot, the disc is raised and pushed over by the other arms. In clambering over rocks or in seaweed and hydroid colonies, the ends of the supple arms often coil about objects. A few ophiuroids reportedly are able to swim, using the same method of progression as in crawling.

There are burrowing, notably members of the family Amphiuridae (*Amphiura* and *Amphipholis*). Muscus-lines burrows with tubelike channels to the surface are excavated in mud or sand using undulatory waves of the arms and digging movements of the tube-feet. The latter swing laterally, moving substrate from beneath the body. The animals never leave the burrows unless dislodged, although the arms may be projected to the surface. Arm undulation provides for burrow ventilation.

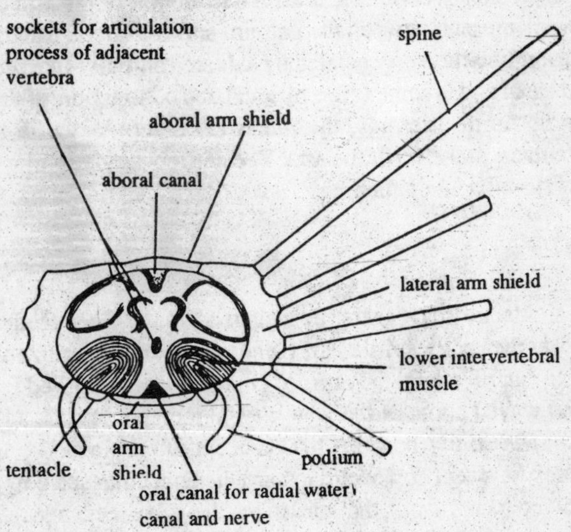

Fig. 4.5. *Ophiocoma*. End of an arm joint

The European *Ophiothrix fragilis* responds to the presence of other individuals and in favourable currents on muddy gravel bottoms attains densities as great as 1000 to 2000 per square meter. Some echinologists consider the ophiuroids to be the most successful group of living echinoderms, they attribute this success in part to their motility, small size, and ability to utilize the protective cover of crevices, holes, spaces beneath stones, and other natural retreats

Water-vascular system

The oral shield that forms the madreporite in ophiuroids usually bears but a single pore and canal. In addition to the oral position of the *madreporite*, the *stone canal* ascends to the water ring, which is located in a groove on the aboral surface of the jaws. The *water ring* bears four *polian vesicles* and also gives rise to the radial canals, which penetrate, through the lower side of the vertebral ossicles of the arms. In each ossicle, the radial canal gives rise to a pair of *lateral canals*, which may extend directly to the pair of ventrolateral podia or may first loop upward and then downward to the podia. The paired lateral canals of ophiuroids contrast with the staggered arrangement of other echinoderms. Ampullae are absent, probabaly correlated with the reduction of the arm coelom, but a valve is present between the podium and the lateral canals. Fluid pressure for protraction is generated by a dilated ampulla-like section of the podial canal and in some forms by localized contraction of the radial water canal. As in asteroids, the radial canal terminates in a small external tentacle at the tip of the arm. The entire water-vascular system is lines with a ciliated epithelium.

Digestive System

Ophiuroids are scavengers, deposit feeders, or filter feeders, and some, such as the Pacific brittle star, *Ophiocomina nigra,* utilize all three methods. In filter feeding, the arms of *Ophiocomina* are lifted from the bottom and waved about in the water. Plankton and detritus adhere to mucous strands strung between the adjacent arm spines.The trapped particles may be swept downward toward the tentacular scale by ciliary currents or collected from the spines by the tube feet, which extent upward for this purpose. The tube feet are then scrapped across the tentacular scales, depositing collected particles in front of the scale. This is also where the ciliary tracts deposit their material. From this point on each side, the food particles are picked up by adjacent podia, compacted into a bolus, and passed along the midoral line of the arm toward the mouth. The food balls are moved by the podia until they reach the proximal parts of the arm, in which movement toward the mouth is facilitated by cilia.

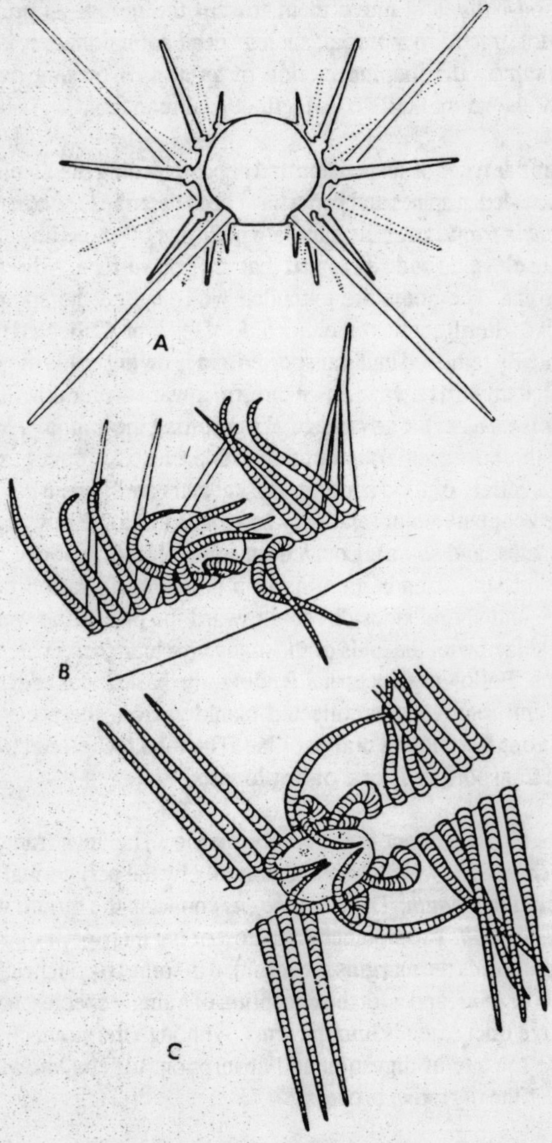

Fig.4.6. *Ophiothrix*. Showing filter feeding. A = End view of arm section showing arm and position, B = Orally, C = Particles are added to a bolons

Deposit feeding on intermediate size particles performed by the podia. The podia collect the particles from the substratum, compact them into food balls, and move them toward the mouth as previously described. Large food material, such as dead animal matter, is swept into the mouth by the looping motion of an arm. Browsing over algae or carrion, the animal utilize its teeth or oral tube feet.

Similar types of suspension and deposit feeding mechanisms involving mucus-laden spines and podia have been described for other ophiuroids. *Ophiothrix fragilis* uses its papillate podia for filter feeding. The feeding arms are elevated and so twisted that the oral surface is directed toward the current. The podia are extended well beyond the spines, forming comblike filtering series on either side of the arm. Collected particles are periodically removed and transported as a growing bolus by action of the podia that travel as a wave down the arm toward the mouth. Such indirect mechanisms have the advantage of permitting the animal to extend only two or three feeding arms out from its protective retreat as well as to utilize a veriety of food sources. Basket stars are suspension feeders but capture zooplankton of relatively large size (10 to 30 mm - crustaceans, polychaetes, and so on). Located on some elevation above the bottom, the basket star extends its arms in a parabolic filtration fan with the concave and aboral sides directed toward the prevailing water current. Prey is seized with the ends of the many arm branches, which coil about the catch. Following nocturnal feeding, the basket star retreats to some shelter and removes the collected plankton from the arms by passing them through comblike oral papillae. The tube feet do not play the role of food transport, as is true of brittle stars.

The digestive tract is extremely simple . The jaws frame a shallow *prebuccal cavity*, which is roofed aborally by the peristomial membrane containing the *mouth.* The *oesophagus* connects the mouth with a large saclike *stomach.* The stomach fills most of the interior of the disc, and in most ophiuroids the margins are infolded to form 10 pouches. There is no *intestine* or *anus*, and with the exception of a single species, no part of the digesative tract extends into the arms. Although the stomach must obviously be the site of digestion and absorption, little is known about the details of the digestive process.

Circulatory System

The coelom in ophiuroids is much reduced when compared with that of other echinoderms. The vertebral ossicles restrict the coelom to the abroal part of the arms the stomach, bursae and gonads leave only small coelomic spaces in the disc. The hemal system is essentially like that of asteriods.

Respiration and Excretion

Gas exchange in ophiuroids takes place by means of 10 internal sacs (bursae) that represent invaginations of the oral surface of the disc. Internally each bursa is lodged between two stomach pouches. The bursae are connected to the outside by way of a slit that runs along the margins of the arms on the oral surface of the disc. Commonly the two openings. The bursae may be lined with ciliated epithelium, especially the slits. The beating cilia create a current of water that enters the peripheral end of the slit, passes through the bursa, and flows out the oral end of the slit.

Many species also pump water into and out of the bursae by raising and lowering the oral or aboral disc wall or by contraction of certain disc muscles associated with the bursae.

The thin-walled respiratory bursae may well be the principal center for waste removal, including waste-laden coelomocytes.

Nervous system

The nervous system is doubled and is composed of the outer sensory and motor ectoneural system and the inner motor hyponeural system. The two systems parallel each other and are separated only by a thin layer of connective tissue. As in asteroids, they are composed of a circumoral nerve ring and radial nerves. Each arm contains a nervous center, and arm dominance and coordination of locomotion are dependent upon interaction of these centers. There are no specialized sense organs in ophiiuroids, the general epithelial sensory cells composing the sensory system. Ophiuroids are negatively phototropic, as are many asteroids. They are also able to detect food without contact. The chemoreceptors are perhaps located on the podia

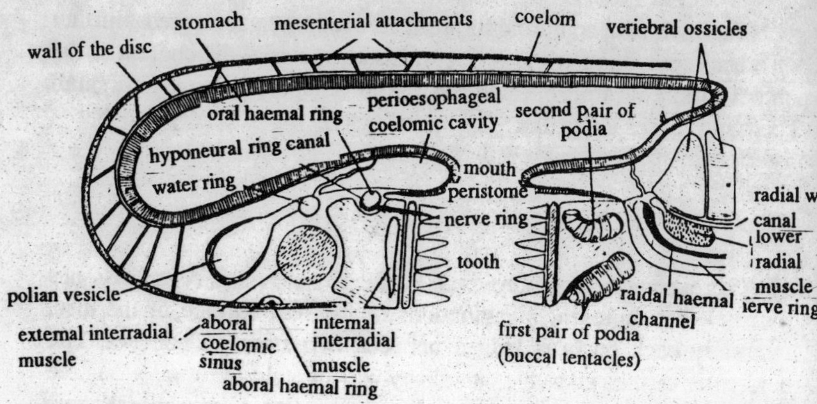

Fig. 4.7. A diagrammatic V.S. of disc, passing through an interradius on left and a radius with arm base on right.

Regeneration

Many ophiuroids can cast off, or autotomize, one or more arms if disturbed or seized by a predator. A break can occur at any point beyond the disc, the lost portion is then regenerated. In most ophiuroids, the disc does not have the regenerative powers of the arms, and in at least some species the entire disc and at least one arm must be present for the *remaining arms to regenerate*.

Asexual Reproduction

However, there are some ophiuroids, notably the six-armed *ophiactis,* in which asexual reproduction by fissiparity takes place. The disc divides into two pieces, each piece with three arms. Fission can take place along any plane, the missing half is then regenerated.

Reproduction

The majority of ophiuroids are *dioecious*. They usually exhibit no sexual dimorphism, however, the females of three species are known to carry dwarf males about, the two sexes being attached mouth to mouth. The gonads are small sacs attached to the coelomic side of the bursae near the bursal slit. There may be one, two, or numerous gonads per bursa with various positions of attachment. Hermaphroditic species are not uncommon. Some bear separate testis and ovaries; others are *protandric*. When the gonads are ripe, they discharge into the bursae, probably by rupture, and the sex cells are carries out of the body in the respiratory water current. Fertilization and development take place in the sea water in many species, but brooding is common. The bursae are used as brood chambers, and development takes place within the mother until the juvenile stage is reached. In most species, only a few young are brooded in each bursae. In some brooding brittle stars, such as *Amphipholis squamata*, development is *viviparous*, for the embryo is attached to the bursal wall and receives nourishment from the mother. In a few species, such as the Antarctic *Ophionotus hexactis*, a single egg is formed and passes into an ovarian lumen, in which both fertilization and development take place, the young serpent star ruptures into the bursa.

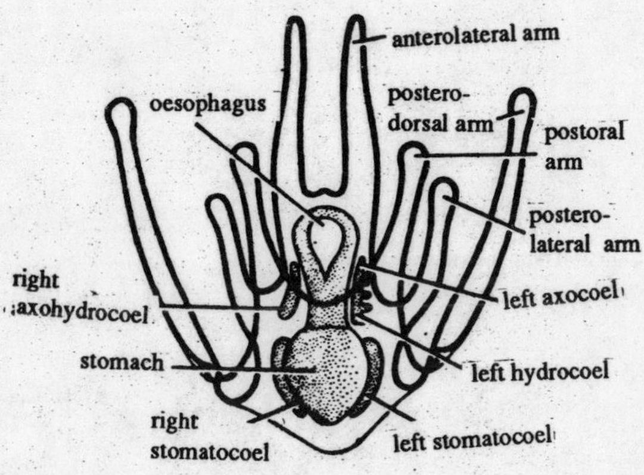

Fig. 4.8. Ophioplutes

Development

In nonbrodding oviparous ohiuroids, early development is similar to that in the asteroids. The larva of many species, called an *ophiopluteus*, displays four pairs of elongate arms supported by calcareous rods and bearing ciliated bands. The shape of the larva is distinctive and quite different from that of the brachiolaria of asteroids. Metamorphosis takes place while the larva is still free-swimming, and there is no attachment stage. The tiny serpent star sinks to the bottom and takes up an adult existence. For the some six species for which information is available, development to this point takes 14 to 40 days. To reach the same shape in the single brooding species studied takes three to seven months. Of the 2000 described species of ophiuroids, larvae are known for 71, and 55 are known to brood. The developmental patterns of the remainder are still unknown.

5

SEA URCHINS

SYSTEMATIC POSITION

Phylum	-	Echinodermata
Subphylum	-	Eleutherozoa
Class	-	Echinoidea
Subclass	-	Regularia
Order	-	Diadematoidea
Suborder	-	Camarodonta

The echinoids are free-moving echinoderms commonly known as *sea urchins, heart urchins,* and *sand doller*. About 900 species have been described. The name Echinoidea, which means "like a hedgehog," is derived from the fact that the bodies of these animals are covered by a shall composed of usually closely fitting calcareous, pilates beset with movable spines.

The echinoid body does not possess arms like that of asteroids, yet they have the same fundamental structure. Rather the shape is circular or oval, and the body is spherical or greatly flattened along the oral-aboral axis. Both tube feet and spires are used in locomotion. The class is particularly intresting from the standpoint of symmetrical for although the sea urchines are radially symmetrical, many soft bottom-dwelling members of the class display various stages in the attaiment of a secondary bilateral symmetry. A third distictive feature of echinoid structure is the flatten and suturing of the skeletal ossicles into a soild case (the test). Fishes, sea stares and other marine carnivores are their enemies.

External Structure

Regular Echinoids. The radial, or regular, members of the class are known as sea urchins. In these forms, body is more or less spherical in shape and armed with relatively long movable spines. Sea urchins are brown, black, purple, green, white, and red, and some are multicolored. Most sea urchins are 6 to 12 cm. in diameter, but some Indo-Pacific species may attain a diameter of nearly 36 cm.

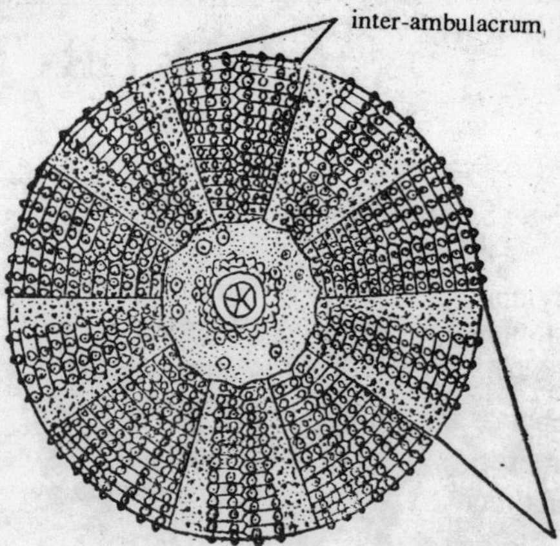

Fig. 5.1. *Echinus*. Shell in oral view

The sea urchin body can be divided into an aboral and an oral hemisphere, with the parts arranged raidally around the polar axis. The oral pole bears the *mouth* and is directed against the substratum. The mouth is surrounded by a peristomial membrane that is thickened along the inner edge to form a lip. The peristome of sea urchins contains small embeded supporting ossicles, and the surface bears a number of different structures arranged in a radial manner. There are five pairs of short, heavy modified podia, called *buccal podia*, and five pairs of bushy projections, called *gills*. In addition to the buccal podia and the gills, the area around the peristome bears small spines and pedicellariae. The aboral pole contains the anal region, known as the *periporct*. The periproct is a small circular membrane containing the *anus*, usually in the center, and a varying number of embedded plates. In the Arbaciidae, there are four large periproct plates, but in other families of sea urchins the plates are smaller and more numerous. The globose body surface can be divided into 10 radial section extending between the oral and aboral poles. Five sections contain tube feet and are called *ambulacral* areas. Between the ambulacral areas are section devoid of podia, known as *interambulacral areas.* The body surface is thus composed of five ambulacral areas. Alternating with five interambulacral areas, all 10 areas converging at the two poles. The movable spines, which are so characteristic of sea urchins, are arranged moe or less symmetrically in the ambulacral and interambulacral areas. The spines are longest around the equator and shortest at the poles. Most sea urchins possess long (primary) and short (secondary) types of spines, the two types being more or less equally distributed over the body surface. However, *Arbacia punctulata*, the common Atlantic sea urchin, possesses only the long type.

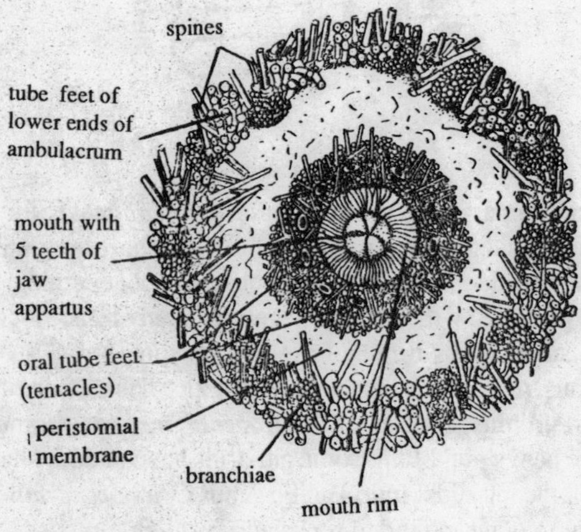

Fig. 5.2. *Echinus*. A aboral view.

Each spine contains a concave socket at the base that fits over a corresponding tubercles on the test. Two sheaths of muscle fibres that extend between the spine base and the test encircle the ball-and-socket joint. Contractions of the outer muscular sheath serve to incline the spines in one direction or another. The inner sheath of muscle fibres (the cog muscle) by uniform contraction of its fibers causes a rigid erection of the spines. The spines are usually cylindrical and taper a point, but

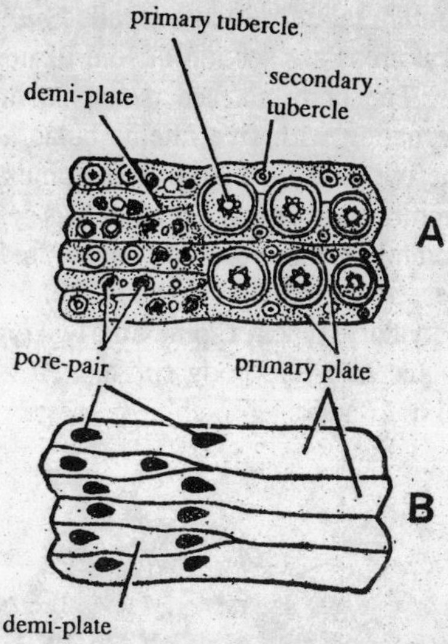

Fig. 5.3. Two ambulacral plates

many species depart from this generalization. *Diadema* of tropical reefs has very long needle-like spines, which can be tilted and waved in direction of intruders. The spines, which can be regenerated, are hollow, brittle, and provided with an irritant, and the outer surface is covered with circlets of small barbs directed toward the spine tip. This urchin can inflict serious, painful wounds if stepped upon. The primary spines of the slate-pencil urchins, species of *Heterocentrotus,* and some species of cidarids are heavy and blunt; some cidiarids have paddle-shaped spines. Members of the intertidal Indo-Pacific genus *Colobocentrotus* are adapted for clinging to wave-pounded rocks and surviving exposue at low tide. The aboral spines are short and heavy, and the blunt tips are polygonal in cross section. These spines fit together like tiles, providing an effective water-shedding surface and protection against desiccation. The Indo-Pacific sea urchin, *Asthenosoma,* bears special poison spines on the aboral surface. The tip of each spine is surrounded by a large blue sac containing poison secreted by the epithelial lining. The poison is highly toxic and painful to man. This genus is also unusual in that it has articulating test plates. When the urchin is taken out of the water the test collapses.

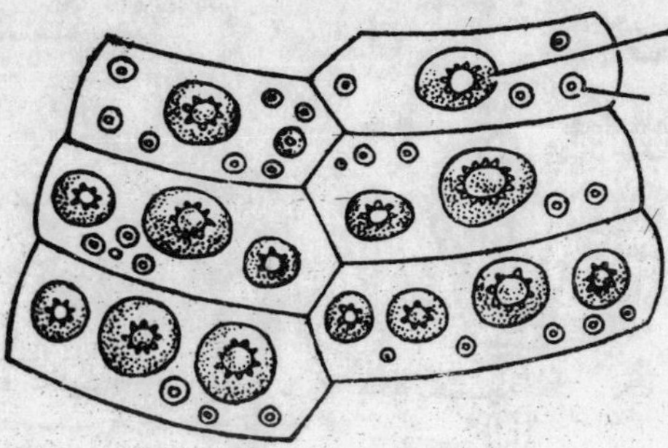

Fig. 5.4. Interambulacral plates.

Pedicellariae, which are characteristic of all echinoids, are located over the general body surface as well as on the peristome. The echinoid pedicellariae are somewhat analogous to the stalked pedicellariae of the asteroids, in that each consist of a long stalk surmounted by jaws. However, in echinoid pedicellariae, the stalk may contain a supporting skeletal rod, and there are usually three opposing jaws. The bases of the jaw ossicles articulate against one another, and not against a basal ossicle as in the asteroids. Muscles at the base of the stalk provide for elevation and direction of the pedicellariae in response to certain stinili.

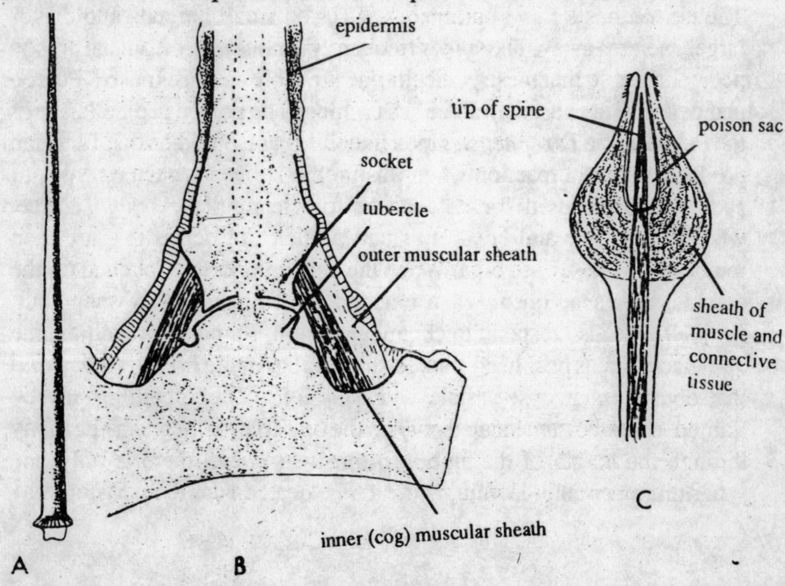

Fig. 5.5. Spines. A = Primary Spine, B = Section through the base of spine
C = Poison spine

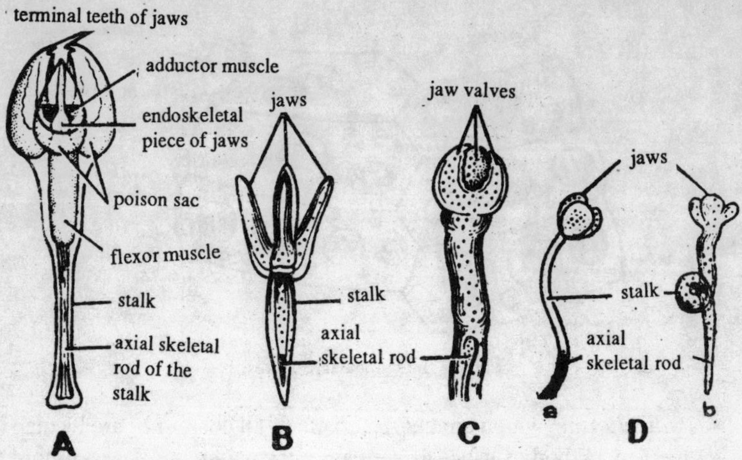

Fig. 5.6. Different types of pedicellarirae of sa-urchin.
A--Gemmiform pedicellaria; B-Tridactyle pedicellaria;
C--Ophiocephalous pedicellaria; D--Trifoliate pedicellaria;
a-Trifoliate pedicellaria of *Trifpneustes*, b--*pedicellaria of Echfnus.*

One of the various types of echinoid pedicellariae, called *globiferous pedicellariae,* contains poison glands. The outer side of each jaw is surrounded by one or two ducts just below the terminal tooth of the jaw. The poison has a rapid paralyzing effect on small animals and drives larger enemies away. The spines frequently incline away from the poison pedicellariae so that these pedicellariae are more easily exposed. Poison pedicellariae are absent in some sea urchins. The poison pedicellariae of the Indo-Pacific *Toxopneustes* look like little parasols when open and can produce a painful reaction when this urchin is handled. Other types of pedicellarie are used for defence or for cleaning the body surface, wherein they bite and break up small particles of debris that are then removed by the surface cilia. When the pedicellariae are touched on the outside, they snap open; when touched on the inside, they snap shut. Pedicellariae also respond to chemical stimuli. In most sea urchins, the ambulacral areas bear head stalked spherical or ovoid bodies (spheridia) that obtain a statocyst vesicle with statoliths. The spheridia may be limited in number and located only on the oral side, or there may be many through the length of the ambulacrum. In *Arbacia,* there is only one spheridium per ambulacrum, and each is located near the peristome.

Body Wall

The body wall of echinoids is composed of the same layers as that in asteroids. A ciliated *epidermis* covers the outer surface, including the spines. Beneath the epidermis lie a nervous layer and then a connective tissue *dermis* the skeletion. The skeleton is composed of ossicles formed into flattened plates that are sutured together to produce a solid immovable test.

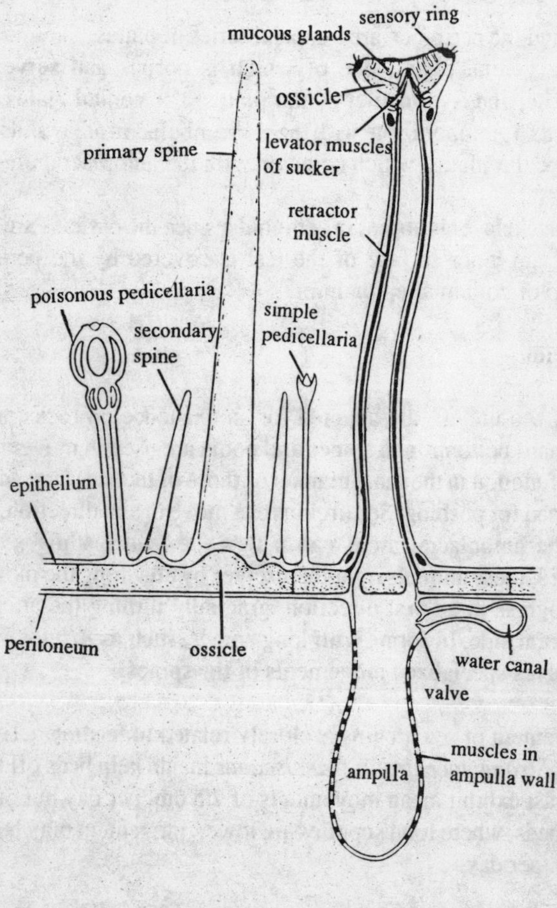

Fig. 5.7. Diagramatic section through the body wall.

In all echinoids, the plates are arranged in rows running from the oral pole to the aboral pole. Each ambulacral area is composed of two rows of ambulacral plates, and each intermbulacral area is composed of two rows interambulacral plates. There are thus 20 rows of plates - 10 ambulacral and 10 interambulacral by holes enabling the canals to connect with the ampullae and podia.

Each of the plates, both ambulacral and interambulacral, bears rounded tubercles on which the spines articulate.

Around the periproct are a special series of plates. These consist of five large genital plates, one of which is porous and serves as the madreporite, and five smaller ocular plates. The genital plates, each of which bears a gonopore, line with the interambulacral areas and alternate with the ocular plates, which coincide with the ambulacral areas.

A muscle layer is absent in echinoids, since the ossicles are immovable, and the inner surface of the test is covered by the peritoneum, composed of columnar epithelium.

Locomotion

Most sea urchins are adapted for an existence on rocks and other types of hard bottoms, and spines and podia are used in movement. The tube feet function in the same manner as those of the sea stars, and spines may be used for pushing. Sea urchins can move in any direction, and any one of the ambulacral areas can act as the lead leading section. If overturned, these animals right themselves by attaching the more aboral podia progress in an oral direction, gradually turning the animal over onto the oral side. In forms with long spines, such as *Arbacia,* righting also involves specialized movements of the spines.

Movement of sea urchins is closely related to feeding activity. For example, *Strongylocentrotus franciscanus* inside kelp beds off the California coast exhibit mean movements of 7.5 cm. per day but outside of the kelp beds, where food supplies are lower, movement may be as great as 50 cm. per day.

Some sea urchins tend to seek rock depressions, and some species are actually capable of increasing the depth of such depressions or even of excavating burrows in rock and other firm material. Burrowing is performed largely by the scraping action of the chewing appartus. Burrowing behavior appears to be an adoption to counteract excessive wave action, and these species are largely found in habitats that are exposed to rough water. One of the most notable burrowing sea urchins is *Paracentrous lividus,* which lives along the coast of Europe. This sea urchin literally riddle rock walls with burrows. When the burrows are shallow, the animal leaves to feed, but it remains permanently within deeper burrows, which often have entrances too small to permit exit. Echinometrids are common burrowing species on tropical reefs. The Indo-Pacific echinometrid *Echnostrephs molars* sits at the mouth of its burrow but drops inside when disturbed. Within the burrow it wedges itself so tightly with its spines that it is virtually impossible to remove. The West Indian *Echinometra* honeycombs coralline rock in surge areas but nonburrowing populations of this species are sometimes encountered. The same is true for *Paracentrotus lividus. Strongylocentrotus purpuratus* is a surge-loving sea urchin that is found along the Pacific coast of North America and commonly burrows where the rock is soft.

Water-vascular System

The water-vascular system of echinoids is essentially like that of the sea stars. One of the genital plates around the periproct contain pores and pore canals and functions as the madreporite. The madreporite, like that of the starfishe, is perforated and opens into a madreporite ampulla lying below it. From this a stone canal decents orally to the water ring, which lies above the peristome in heart urchins or just above the viewing appartus in regular urchins and sand dollars. The radial canals extend from the water ring and run along the underside of the ambulacral areas of the test.

Each radial canal terminates in a small protrusion called a *terminal tentacle,* which penetrates the most apical ambulcral plate. The radial canal gives off alternately on both sides lateral canals to the bases of the ampullae. The canals connecting the ampullae and penetrate the ambulacral ossicles rather than pass between them. Moreover, these canals are also peculiar in being doubled - that is, from each ampulla, two canals pierce the ambulacral plate and become confluent on the outer surface to enter a single podium. The suckers of the podia of sea urchins are highly developed and are provided with a system of muscles and supporting ossicles.

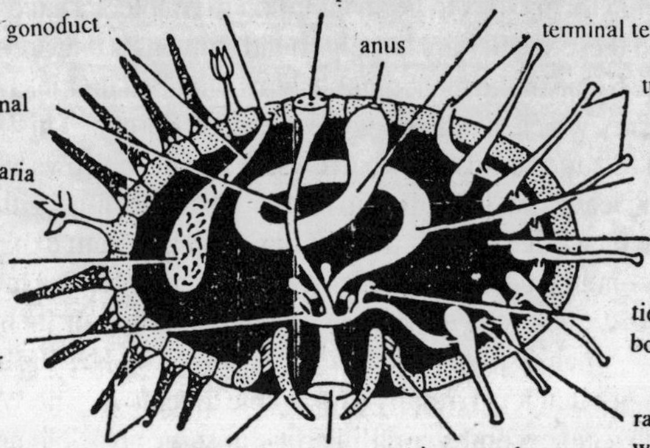

Fig. 5.8. *Echinus*. Showing water vascular System.

Digestive System.

The food of the sea uschins consists of marine vegetables and animals matter which falls to the sea bottom.

Sea urchins feed with highly developed scraping appartus called *Aristotle' lantern.* The appartus is composed of five large calcareous plates called *pyramids*, each of which is shaped somewhat like a barbed arrowhead with with the point projected toward the mouth. The pyramids are arranged radially, with each side connected to that of the adjacent pyramid by means of transverse muscle fibers. Passing down the midline along the inner side of each pyramid is a long calcareous band. The curled upper end of the band is enclosed within a dental sac and is the area of new tooth formation. In *Paracentrotus lividus,* new tooth material is produced at a rate of about 1 to 1.5 mm. per week. The oral end of the band projects beyond the tip of the pyramid as an extremely hard pointed tooth. Since there is one such tooth band for each pyramid, there are five teeth projecting from the oral end of the lantern.

In addition to the teeth and pyramids, the Aristole's lantern is composed of a number of smaller rodlike pieces located at the aboral end. By means of special muscles, the lantern can be partially protracted and retracted through the mouth. Other muscles control the teeth, which can be opened and closed. In the more primitive cidaroids, the scraping

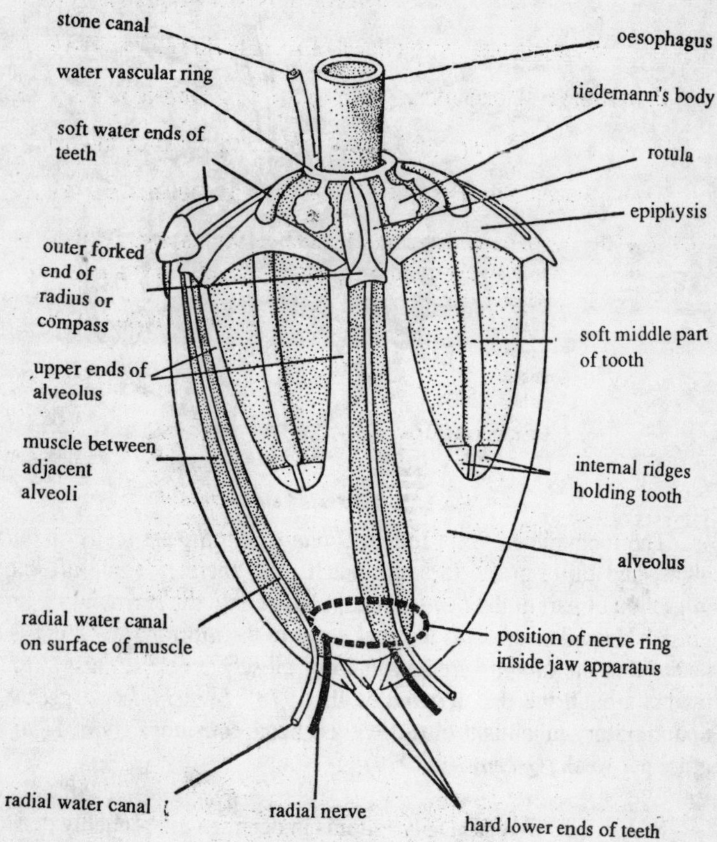

Fig.5.9. Aristotle's lantern.

action of the lantern is largely restricted to opening and closing the teeth, but in higher groups, the lantern can also be swung laterally. The ability of the lantern to be protruded and retracted makes possible pulling and tearing in addition to scraping.

The majority of sea urchins are grazers, scraping the substrate surface on which they live with their teeth. Although algae are usually the most important food, most sea urchins are generalist and include a wide-range of plant and animal material in their diet (*Lawrence,* 1975) Moreover, the diet of a particular species will vary from area to area, depending upon what is available.

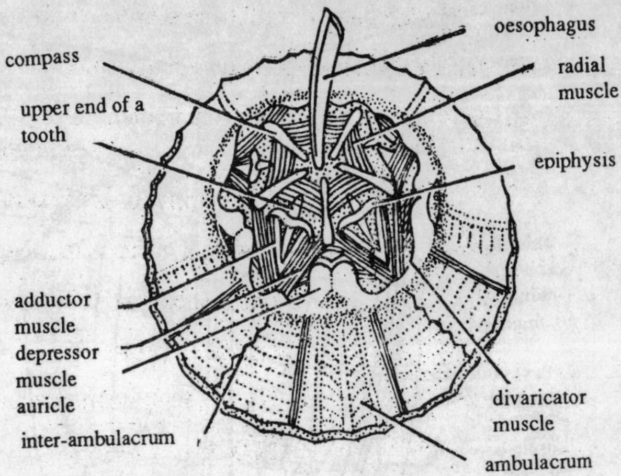

Fig. 5.10. Muscles of Lantern.

The long-spined West Indian *Diadema antillarum* feeds on coral, algae, and turtle grass. In some patch reefs there is a late afternoon migration of part of the *Diadema* population onto the surrounding turtle grass beds and a return to the reef early in the morning. Grazing is so heavy that the urchins create a halo of closely cropped grass 2 to 10 meters around the reef (*Ogden* et al., 1973). *Lyrechinus varegatus,* a nonmigratory inhabitant of turtle grass beds, consumes about 1 gm. of grass per week (*Greenway,* 1976).

Sea urchins living on soft bottoms in deep sea are probably detritus feeders, consuming minute organic particles in the bottom ooze. Some, sich as *Hygrosoma,* feed exclusively on terrestrial or shallow-water plant material carried down to the deep-sea floor.

Burrowing sea urchins fed on encrusting and endolithic algae on the walls of their burrows as well as algal fragments and other organic debris that is washed in. *Echinostrephus molars* of the Indo-Pacific sits at the mouth of its burrow and when debris touches the spines or long podia, the spine tips close and grasp it.

In sea urchins the interior of Artistotle's lantern contains both a *buccal cavity* and a *pharynx* that ascends through the apparatus and passed into an oesophagus. The oesophagus descends along the outer side of the lantern and joins an *intestine* or *stomach* . At the junction of the oesophagus and intestine a blind pouch or cecum is usually present. The intestine is very long and can be divided into a proximal *small intestine* and a distal *large intestine*. The small intestine makes a complete turn around the inner side of the test wall, to which it is suspended. It then passes into the larger aboral intestine, which makes a complete turn in opposite direction. The large intestine then ascends to join the *rectum*, which empties through the *anus* within the periproct.

In most echinoids a narrow tube, called a *siphon*, parallels the intestine for about one half of its length. The ends of the siphon open into the lumen of the intestine.

The intestine appears to be the site of both digestion and absorption, although the cecum appears to be involved in th production of some enzymes. the siphon supposedly functions in the removal of excess water from the food. The products of digestion pass through the intestine wall into the coelom for distribution. Glycogen is one of the principal food reserves and apparently can be stored in any of the body tissues but may be particularly prevalent in the intestine wall.

Haemocoelomic System

Besides the radial canal of the water vascular system, two other canals transverse the *ambulacra*, in relation to the radial nerve trunks. Of these the *epineural canal* runs between the nerve and the test and the *peseudohaemal canal* runs along the inner axial face of the nerve trunk. Both these end blindly at the apex of the test in close proximity to the terminal tentacle and do not open into the circular epineural canal surrounding the pharynx. Little is known about the ontogenic history of the pseudohaemal system. The canals contain fluid similar to that found in the coelom and probably they provide nourishment to the nerve trunks with which they are so intimate.

The *blood vascular system* of *Echinus* and its allies has been subject of much discussion. There are undoubtedly two vessels which run alongside the interior coil of the intestine in the mesentery which supports the latter. The *ventral vessel* runs along the inter axial the inner axial side and the other along the outer side of the intestine, both being connected by numerous lacunae and both have been described as opening into a *circular vessel* and to surround the oesphagus in close proximity to yet quite distinct from, water vascular canal, The presence of these vessels is doubted by many (*Chadwick*).

There is the *axial sinus* with the *axial organs* within. These are similar to those of the sea-stars described earlier. In *Echinus* there is a *perianal sinus.* This is separated by a membrane which connects the anal end of rectum with the surrounding body below this another similar membrane encloses a second and slightly larger *periproctal sinus.* The fluid that fills the body cavity and sinuses has a peculiar musk-like odour and is of pale grayish red colour. It contains a small quantity of albuminoid matter and has an alkaline reaction. Two kinds of corpuscles float in the fluid, small colourless amoebocytes and larger colourless cells containing a number of large, spherical refractive granules. The coelomic fluid coagulates when withdrawn from the test.

Respiration.

In *Echinus* the five pairs of peristomial grills are probably the chief centers of gas exchange. Each gill is a highly branched out pocketing of the body wall and is therefore lined within and without by a ciliated epithelium. The internal cavities of the gills are filled with fluid and are continuous with a separate division of the coelom, called the *peripharyngeal coelom,* that surrounds the lantern. Special ossicles and their associated muscles, composing the aboral part of the lantern, act as a pumping mechanism, changing the pressure within the peripharyngeal coelom by pulling down the partitioning membrane and thereby forcing fluid into and out of the gills. Pumping is not continuous but depends on the need for oxygen, and the appartus is under the control of the nerve ring. The nerve ring is stimulated to emit impulses to the lantern by an increase in acidity resulting from an accumulation of carbon dioxide.

As in other echinoderms, all of the podia contribute to some gas exchange. In most sea urchins the more aboral podia are modified in various ways for this function (*Fenner,* 1973). The podia and ampullae are commonly septate with a two way circulation of fluid through these structures, one current (on the side nearest of the radial canal) flowing toward the base of the podium and the opposite current flowing toward the podial time. The aboral podia may lack podia and on *Arbacia* are flattened.

Nervous System.

The nervous system consists of a ring of fibers and ganglion cells the *nerve ring,* lying around the pharynx close to the mouth. From this arise five *radial nerves,* traversing the entire length of the ambuolacra, upon the inner surfaces of the test, in close ration with radial canal. From the nerve ring nerves are given off to the integument and the walls of pharynx in which they break up to form plexus. Five lamellae of ganglion cells and nerve fibers lie in close proximity to the nerve ring at the points of origin of the radial trunks. From each lamella a pair of large nerve arises. These ascend along the edges of the five alvoli and are supposed to innervate their muscles. This has been described as *deep nervous system.* But for this, the deep and the aboral nervous systems are poorly developed. Organs of special sense do not occur. *Sphaeridia* and the *terminal tentacles* of the water vascular system are the only known sense organs of *Echinus esculentus.* The tube-feet are highly sensitive to external stimuli and those of *E.microtuberculatus* have a tubercle-like thickening of ectoderm close to the sucker disc, beneath which there is a ganglionic mass of nerve tissue.

Echinoids are in general negatively phototropic and tend to seek the shade of crevices in rocks and shells. Some species of sea urchins, such as *Tripneutstes, Lytechinus, Strongylocentrotus*, cover themselves with shell fragments, and other objects using the tube feet. The significance is uncertain but clearly seems to be a light response in some, for *Tripneustes* covers more in the summer than in the winter, and the related *Lytecheus* drops its cover at night and will cover only an experimental band of light crossing the body.

A few sea urchins, notably species of *Diadema*, possess chromatophores, permitting some diurnal color change. However, this function is still poorly understood.

Reproduction

All echinoids are dioecious, and the majority display no sexual dimorphism. In *Echinus* there are five gonads suspended along the inter-ambulacra on the inner side ~~of~~ the test. Each gonad is covered on the outside with coelomic peritoneum and line internally with a germinal epithelium. Between these two layers are found muscles cells and connective tissue. A short gonoduct extends aborally from each gonad and opens through a gonopore located on one of the five genital plates

Sperm and eggs are shed into the sea water by the contraction of the muscle layers of the gonads, and fertilization takes place in the sea water. As in asteroids, spawning in temperate species takes place in the spring and early summer. Brooding is displayed by some cold-water sea urchins. Brooding sea urchins retain their eggs on the peristome or around the periproct and use the spines in holding the eggs in position.

Development

Cleavage is equal, up to the eight-cell stage, after which the blastomeres at the vegetal pole proliferate a number of small micromeres. A typical blastula ensues, displaying a wall of equal-sized cells, with the exception of the small micromeres and somewhat larger macromeres at the vegetal pole. The blastula becomes ciliated and free-swimming within 12 hours after fertilization.

Grastrulation is typical but is preceded by an interior proliferation of cells by the micromeres, which form the mesenchyme. Additional mesenchymal cells are proliferated from the tip of archenteron. The coelom is formed by the separation of the free end of the archenteron. The separated portion then divides into right and left pouches, or lateral

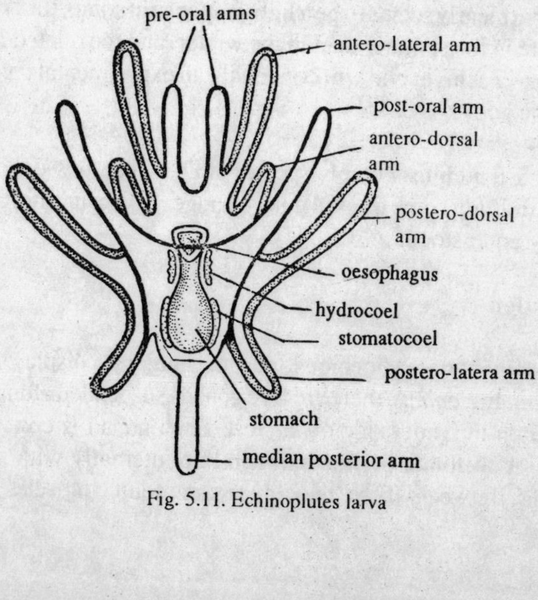

Fig. 5.11. Echinoplutes larva

Fig. 5.12. Juvenile of sea urchin

divisions may appear prior to the separation from the main portion of the archenteron. The gastrula becomes somewhat cone-shaped and gradually develops into a planktonic larval stage, the *echinopluteus,* which bears six pairs of long larval arms and is very similar in appearance to the ophiopluteus of ophiuroids. The close similarity represent concergence, for the larval arms appear to be adaptive for increasing the surface area of the feeding-locomotor bands. The nonfeeding yolk-laden larva (*lecithotrophic)* of some species (*Heliocidaris)* has no arms. The echinopluteus swims and feeds as described for asteroid larvae. Its complete development may take as long as several months. During later larval life, the adult skeleton begins to form, first the five genital plates, then the ocular. the echinopluteus gradually sinks to the bottom; however there is no attachment as in asteroids, and metamorphosis is extremely rapid, taking place in about an hour. Young urchins are no larger than 1 mm. Internal developmental is basically like that of asteroids, although differing in detail.

Juvenile sea urchins may occupy more protective microhabitats than do the adults, such as beneath stones and shells or within kelp. Most *Strongylocentrotus franciscanus* juveniles live beneath the protective oral spines of adults sharing the adults food and moving to the periphery of adults with increased size.

6

SEA CUCUMBERS

The holothuroids (Gr. *holos,* whole + *thourios,* rushing) are a class of some 900 echinoderms known as sea cucumbers. Like echinoids, the body of the holothuroids is not drawn out into arms, and the mouth and anus are located at opposite poles. Also, there are ambulacral and interambulacral areas arranged meridianally around the polar axis. However, holothuroids are distinguished from other echinoderms in having the polar axis greatly lengthened, which results in the body having an elongated cumcumber shape. This shape forces the animal to lie with the side of the body, rather than the oral pole, against the substratum. The class is further distinguished from other echinoderms by the reduction of the from other echinoderms by the reduction of skeletion of microscopic ossicle and by the skeleton to microscopic ossicles and by the modification of the buccal podia into a circle of tentacles around the mouth. Thus their most striking features are the *muscular body wall* almost devoid of large skeletal plates, the *branching tentacle,* surrounding the mouth.

External Feature

Most sea cumcumbers are colored black, brown, or olive green, but there are some rose, orange, and violet species, and striped patterns are not infrequent. There is considerable range in size. The smallest species are less than 3 cm. in length (oral to aboral end), whereas Stichopus from the Philippines may attain a length of 1 meter and a diameter of 23 cm. Most of the common North American and European species, such as *Cumcumaria, Holothuria, Thyone, and Leptosynapta,* range from10 to 30 cm. in length.

The body shape varies from almost spherical, as in Ypsilothuria, to long and wormlike, as in the synaptid sea cumcumbers. Not infrequently, the mouth and anus are displaced from the ends of the long axis anus are displaced from the ends of the long axis of the body. Although there are a few genera, such as *Psolus* and *Ceto,* that are provided with a protective armor of calcareous plates (modified surface ossicles), the body surface in the majority of sea cumcumbers is leathery as the result of reduction of the skeletal ossicles to microscopic size.

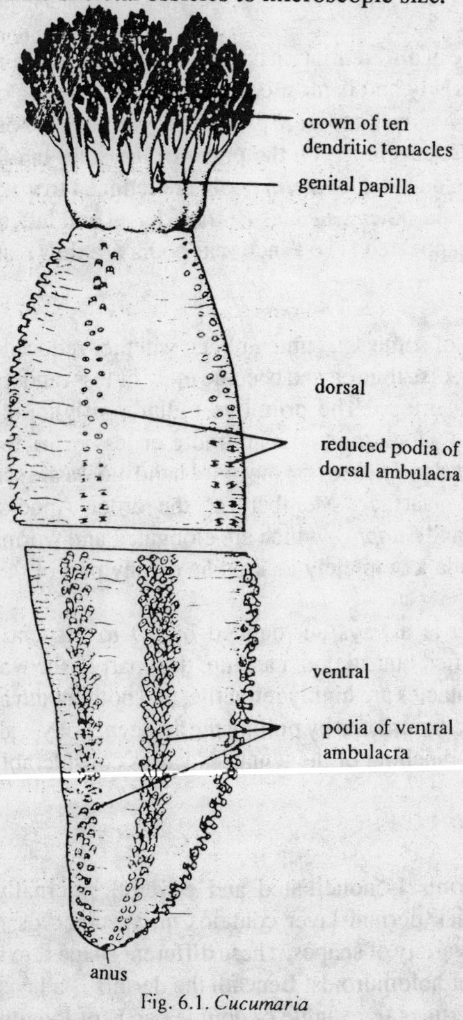

Fig. 6.1. *Cucumaria*

Holothuroids lie with one side of the body against the substratum, and this ventral surface is composed of three ambulacral areas (trivium), commonly called the sole. The dorsal surface consists of two ambulacral areas. As might be expected, there has been a tendency for the dorsal and ventral surfaces to become differentiated, thus producing a secondary bilateral symmetry. Not that the bilateral symmetry of holothuroids has evolved in an entirely different manner from that of the irregular echinoids.

The degree of differentiation between the dorsal and ventral surfaces varies considerably and is most easily seen in the nature of the regular podia. In *Cumcumaria,* podia are present on both surfaces of the body but suckers are best developed on the podia of the sole. In other's such as *Holothuria,* the dorsal and lateral podia are reduced to warts, tubercles, or papillae. In still others, such as *Psolus,* dorsal and lateral podia have completely disappeared. The functional podia are thus restricted to the creeping sole.

The podia of some sea cumcumbers, whether reduced or not, have lost their radial, distribution and become more or less randomly scattered over the body surface. The primitive radial configuration is seen in *Cucumaria,* on which the podia are more or less restricted to the five ambulacral areasl in *Thyone,* on the other hand, podia are scatttered over the entire body surface. Members of the order Apodida (*Synapta, Leptosynapta,* and *Euapta),* which are elongated and wormlike, and the order Molpadiida, completely lack podia on any part of the body.

The mouth is always surrounded by 10 to 30 tentacles, which represent modified buccal podia and are thus part of the water-vascular system. The tentacles are highly retractile, and both mouth and tentacles can be completely retracted by pulling the adjacent body wall over them. The form and branching of the tentacles varies considerably.

Body Wall

The epidermis is nonciliated and covered externally by a thin cuticle. The thick dermal layer contains microscopic ossicles, which display a great variety of shapes. These different shapes are important in the taxonomy of holothuroids. Beneath the dermis is a layer of circular muscle that overlines five single or double bands of longitudinal fibres located in the ambulacral areas. The body wall of sea cucumbers (called trepang) is a culinary delicacy in the Orient. Large species of sea cucumbers are collected and boiled, which causes their bodies to contract and thicken and also brings about evisceration of the internal organs. The body wall is then dried and solde, mostly to Chinese, as trepang or beche-de-mer. Trepang, when cooked in certain dishes, imparts a distinctive flavor to the food.

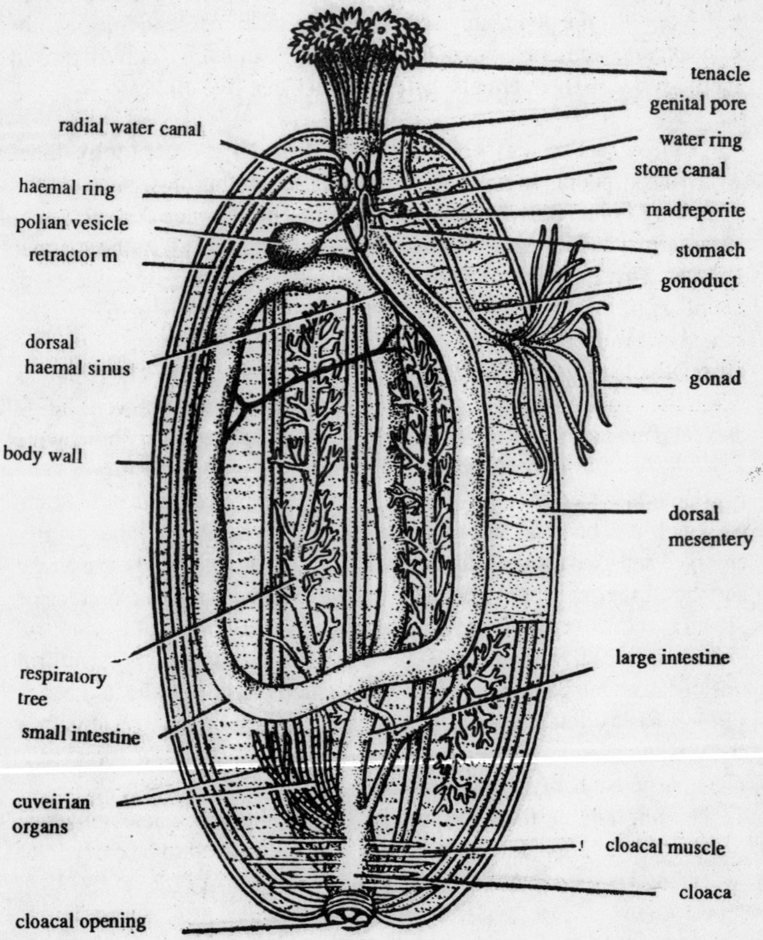

Fig. 6.2. *Holothuria*. Showing in temal anatomy

Locomotion

Sea cucumbers are relatively sluggish animals and live on the bottom surface or burrow in sand and mud. Forms with podia may creep along on the sole, with the podia functioning as those in in asteroids. Righting is accomplished by twisting the oral end around untill the podia contact the substratum. Many hard-bottom forms live beneath stone, and some, such as species of *Holothuria, Cucumaria,* and *Psolus,* are so sedentary that the podia are used more for attachment than for locomotion. Others, such as species of the large, tough *Stichopus,* crawl exposed on the surface. There are also a few species that live on algae.

Burrowing species include the Apodida and ther order Molpadiida, which lack, podia, as well as some pedate holothuroids, such as the common *Thyone*. Burrowing is accomplished by alternate contraction of longitudinal and circular muscle layers of the body wall in the manner employed by earthworms. The tentacles aid by pushing away the sand. members of the Apodida burrow completely beneath the surface. The other burrowers, the Molpadiida and species of *Thyone* and *Cucumaria,* are relatively sedentary and excavate a U-shaqped burrow. The tentacles extend from one opening of the burrow, and the pusating anal ventilating current is maintained through the opposite opening. Sedentary burrowers move very little once they have attainted the proper positionn. The Elasipodida are a curious group of entirely deep-sea holothuroids, some of which are benthic and some pelagic. The podia may be greatly enlarged and used for walking, for most deep-sea holothuroids live on the surface of the sea floor. The sole is quite flattened, and the tentacular crown is turned ventrally, giving them a markedly bilateral symmetry. Pelagic species have papillae webbed together in various ways to form fins or sails. One genus, *Pelagothuria,* which has been collected at the surface, has a circlet of long, webbed papillae just behind the tentacles and probably swims like a jellyfish. *Peniagone* possesses a dorsal sail. The transparent bathypelagic *Peniagone diaphona* lives up to 70 m. off the bottom, holding the body in a vertical position, tentacles upward. About half of the species of the Elasipodida are belived to be capable of swimming.

Feeding and Digestion

Sea cucumbers are chiefly *deposit* or *suspension feeders.* More or less stationary burrowers, such as *Thyone* and *Cucumaria,* or forms that

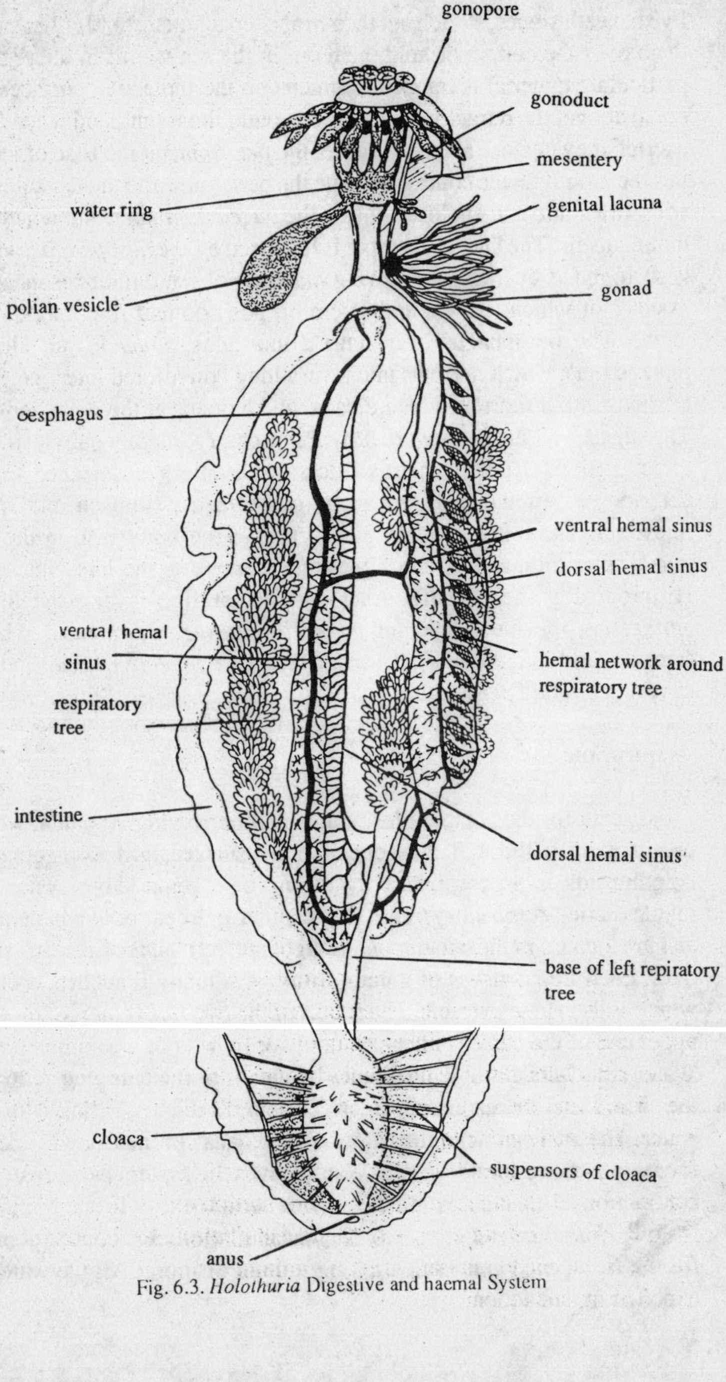

Fig. 6.3. *Holothuria* Digestive and haemal System

live beneath stones, stretch out their branched tentacles and either sweep them over the bottom or hold them out in the sea water. In either case, particulate material is trapped in mucus on the tentacular surfaces. As noted above, the retractile tentacles surround the mouth and when fully exerted they enclose a narrow space, the *peristome* at the base of which lies the mouth. In the contracted state the peristome and the tentacles are withdrawn into and enclosed within the *buccal chamber* into which the mouth opens. The buccal chamber is followed by the *oesophagus,* which is surrounded by a calcareous ring made up of ten circumoesophageal ossicles of which five are ambulacral in position and five inter-ambulacral. The oesophagus leads into a muscular *stomach* (not clearly marked out), which narrows into a very long convoluted intestine. This expands into a short muscular cloaca which opens at the posterior end. The digestive tract is supported by a mesentery, mostly subvided into three portions. The first part of dorsal mesentery is attached to the anterior part of the digestive tract (oesophagus, stomach and small intestine). The left mesentery is attached to the body wall in the left dorsal inter-radius, the ventral mesentery supports the large intestine. Histologically the digestive tract consists of five layers; the lining *epithelium*, the inner layer of *connective tissue*, the muscular layer (*circular* and *longitudinal*), the outer layer of *connective tissue* and the covering *ciliated peritoneum.*

Respiration

Except for the pelagic Elasipodida and burrowing Apodida, which obtain oxygen through the general body surface, gas exchqange in holothuroids is accomplished by means of a remarkable system of tubules called *respiratory trees.* The respiratory trees are two in number and are located in the coelom on the right and left sides of the digestive tract. Each tree consists of a main trunk with many branches, each of which ends in a tiny vesicle. The trunks of the two trees emerge from the upper end of the colaca either separately or by way of a common turnk. Water ciruclates through the tubules by means of the pumping action of the cloaca and the respiratory trees. The cloaca dilates, filling with sea water. The anal sphincter then closes, the cloaca contracts, and water is forces into the respiratory trees. Water leaves the system because of the contraction of the tubules and the reverse action of the cloaca. Pumping is slow; *Holothuria* requires 6 to 10 cloacal dilations and contractions to fill the trees, each contraction taking minute or more. All the water is expelled in one action.

Gases are exchanges with the coelomic fluid and indirectly with the hemal system described later. An interesting commesnsal relationship is that between the slender tropical pearlfish and sea cucumbers. These little fish, which are about 15 cm. long, make their home in the trunk of one of the respiratory trees of certain sea cucumbers.

The fish leaves the host at night while it searches for food, after such excursions, the fish forces its way into the anus and cloaca and back to the shelter of the respiratory tree.

The apodida, which lack respiratory trees, possess peculiar coelomic organs (ciliated funnels or ciliated urns) that probabaly have an excretory function. The urns are attached to the bases of the mesenteries or to the coelomic wall near the attachments of the mesenteries. They are relatively numerous and are sometimes clustered together. Each ciliated urn is covered externally with pertitoneum and is lined internally with ciliated columnar epithelium. Coelomocytes laden with waste enter the funnels and then pass into the body wall.

Haemal System

Holothuroids, especially the Aspidochirotida (*Holothuria* and *Stichopus)*, also possess the most highly developed hemal system of any of the echinoderm classes. The general organization of the system is essentially like that of other echinoderms. A *hemal ring* and *radial hemal sinuses* parallel the water ring and the radial canals of the water-vascular system. The most conspicuous features of the system, at least in larger species, are a dorsal and a ventral vessel that accompany the intestine. These main intestinal sinuses supply the intestinal wall with a large number of smaller channels. In the region of the descending small intestine at least in *Isostichopus badionotus* 120 to 150 single-chambered hearts pump blood from the dorsal vessel through a system of intestinal lamellae that project into the intestinal lumen. From the intestinal lamellae blood collects in the primary ventral vessel. The left respiratory tree is intimately associated with the hemal system of the ascending small intestine. The tree lies within an enveloping mesh of tiny follicular and straight vessels that drain into a secondary dorsal vessel.

Most of the hemal vessels are well developed, containing a muscle and a connective tissue layer, and covered on the outside with ciliated pertitoneum and on the inside with endothelium The blood is essentially like the coelomic fluid, in fact, coelomocytes are formed in the walls of certain vessels. Peristaltic contractions of the dorsal vessel are of primary importance in propelling blood, but although circulation is largely unidirectional, it is not rapid, and there is some ebb and flow. The holothuroid hemal system is clearly involved in some gas transport and appears to play some role in absorption or food transport, but the exact functional significance remains to be determined.

Exeretion

Most ammonia probabaly exists by diffusion through the respiratory trees. Particulate waste, as well as nitrogenous material in crystalline form, is carried by coelomocytes from various parts of the body to the gonadal tubules, the respiratory tree, and the intestine. Waste then leaves the body by way of the lumina of these organs.

Water-Vascular System

Although the water-vascular system of holothuroids is basically like that of other echinoderms, the *madreporite* in most species is peculiar in

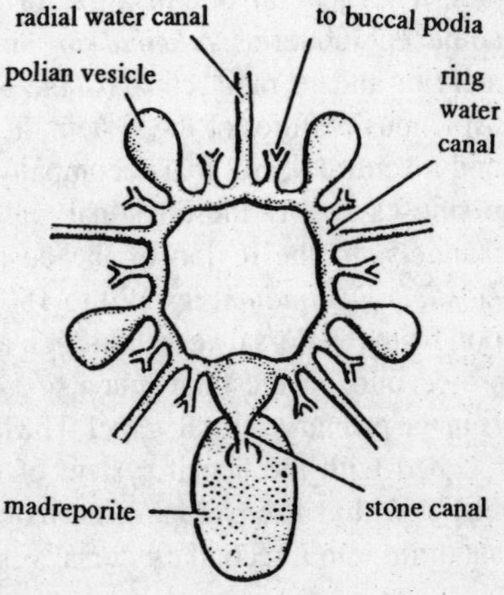

Fig. 6.4. *Cucumaria*. Water vascular organs.

having lost connection with the body surface and in being unattached in the coelom. Pores and pore canals are still present in the madreporite, but coelomic fluid rather than sea water enters and leaves the system. The madreporite hangs just beneath the base of the pharynx and is connected to the water ring by a short *stone canal*. The *water ring* encircles the base of the pharynx and gives rise to *polian vesicles,* which hang into the coelom. The number of polian vesicles varies considerably. There may be 1 *(Cucumaria)*, 3, 4 (*Thyone*), or as many as 10 or even 50 in certain Apodida. The vesicles are believed to function as expansion chambers in maintaining pressure within the water-vascular system.

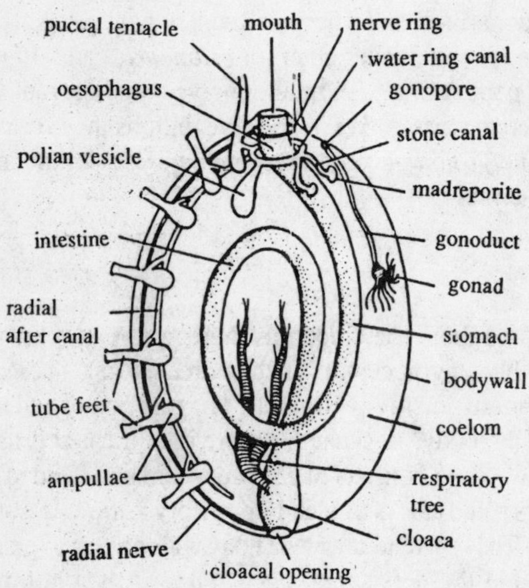

Fig. 6.5. Diagrammatic L.S. of holothurian showing water Vascular System

From the water ring, five *radial canals* pass upward to the inner side of the calcareous ring and then outward through a notch at the end of each radial plate. Just before leaving the calcareous ring, each radial canal gives off smaller canals to the tentacles. On leaving the ring, the radial canals then pass posteriorly within the body wall along the length of the ambulacra. Here *lateral canals* supply the podia. Ampullae are present for both podia and tentacles, although when the podia are reduced, there is a corresponding reduction in the ampullae. In the Apodida, which lack tube-feet the water-vascular system is limited to the oral water ring, the polian vesicles, and the buccal podia (tentacles).

Nervous System

The nervous system consists of the main nerve ring, the *radial nerves* and their branches. The nerve ring is a circular or somewhat pentagonal flattened band situated in the buccal membrance close to the bases of the tentacles. From the outer surface the nerve ring gives a strong ganglionated nerve into each tentacle. Nerve supply to buccal membrane and pharynx also arises from the inner surface of the nerve ring. A *radial nerve* passes into each radius. There are such five radial nerves coursing along the ambulacra. Each radial nerve is a somewhat flattened ganglionated cord dividing into a thicker outer (or *ectoneural*) and thinner inner (or *hyponeural*) part. Along with smaller nerves arising from them and innervating different parts of their respective regions they form epieural and hyponeural nerve complexes corresponding with similar complexes in other echinoderms.

Sense organs

Scattered throughout the epidermis there are numerous neurosensory cells (probably tangoreceptors or chemoreceptors). These are more numerous in the more sensitive body ends. They are connected to general nerve-plexus of the body. In synaptids more complicated sense organs occur. Each sense organ consists of a sensory bud surrounded by gland cell. The sense organ, thus, is of glandulosensory nature. In some statocysts are present. They serve to orient the body with respect to gravity.They are situated around the nerve ring. Each statocyst is a hollow sphere enclosing 1-20 vacuolated cells containing a refringent inorganic material named lithocytes. The general body surface of holothurians is sensitive to light but definite sense organs responsible for this have not been indetified yet. *Holothuria* retracts its tentacles when these are locally illuminated, and reacts to general illumination by moving and turning.

Regeneration

The expulsion of sticky tubules from the anal region is commonly associated with sea cucumbers, but this defensive phenomenon is actu-

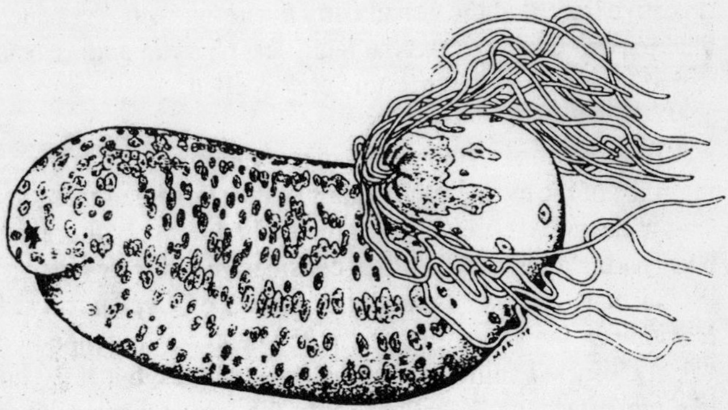

Fig. 6.6. *Holothuria*. tubules of Cuvier.

ally limited to some species of the genera *Holothuria* and *Actinopyga*. Such sea cucumbers possess from a few to a large mass of white, pink, or red blind tubules (tubules of *Cuvier*) attached to the base of one (frequently the left) or both of the respiratory trees or to the common trunk of the two trees. When these sea cucumbers are irrittated or attacked by some predator, the anus is directed toward the intruder, the body wall contracts, and by ruptture of the cloaca the tubules are shot out of the anus.

The cuvierian tubules of some species are not adhesive but liberate a toxic substance, holothurin (a saponin). Holothurin is also found in the body wall of some species. South Pacific islanders have long used the macerated bodies of certain sea cucumbers to catch tide pool fish. During the process of expulsion, each tubule is greatly elongated by water forces into its lumen, and the tubules break free from their attachment to the respiratory tree. In *Holothuria*, the detached tubules are sticky and entangle the intruder in a mesh of adhensive threads. Small crabs and lobsters may be rendered completely helpless and left to die slowly, while the sea cucumber crawls away. After discharge, the tubules of *Cuvier* are regenerated. Often confused with the discharge of the tubules of Cuvier is a more common phenomenon (called *evisceration*) that occurs in many holothuroids. Evisceration in the case of some genera, such as *Holothuria, Stichopus,* and *Actinopyga,* involves the rupture of the cloaca and the expulsion of one or both respiratory trees, the digestive tract, and the gonads. In *Thyone* and other sea cucumbers, the anterior end ruptures, and the tentacles, pharynx and associated organs and at least part of the intestine are expelled.

Evisceration is followed by regeneration of the lost part; the remaining stubs of the eviscerated organs or the associated mesenteries are the sites of the initial regenerative growth. In the case of more radical loss of body parts, holothuroids differ considerably in their regenerative powers. In many forms, particularly species of *Cucumaria, Thyone, and Holothuria,* the cloacal region is the center of regeneration. In transversely bisected animals, each half regenerates, but if the animal is cut into numerous small sections, only the terminal piece, containing the cloaca, regenerates. In the burrowing synaptids, on the other hand, only the section of the body containing the anterior end is capable of regeneration, and all posterior pieces eventually die.

Reproduction

Holothuroids differ from all other living echinoderms in possessing a single gonad, a condition that is believed to be primitive. Most cucumbers are *dioecious*, and the gonad is located anteriorly in the coelom beneath the mid-dorsal interambulacrum. There are a few protandric hermaphrodities. Typically the gonad is composed of a large cluster of simple or branched tubules joining together at the end of the gonoduct and attached to the left side of the dorsal mesentery suspending the anterior region of the gut. The gonoduct plus the tubules thus resemble a mop. Not infrequently, the tubules are divided into two clusters, one on each side of the dorsal mesentery, in the synaptids, the gonad is composed of only two branched tubules. The *gonopore* is located mid-dorsally between the bases of two tentacles or just behind the tentacular collar.

Some 30 brooding species are known, over half of which are cold-water forms, largely Antarctic. During spawning, the eggs are caught by the tentacles and transferred to the sole or to the dorsal body surface for incubation. Frequently these two areas contain special brooding pockets for retaining the eggs; in a few species, such pockets have become so invaginated that they are actually internal. Even more remarkable is coelomic incubation, which takes place in the Californian *Thyone rubra*, *Leptosynapta* from the North Sea, and in a few species from other parts of the world. The eggs pass from the gonads into the coelom and are fertilized in an undiscovered manner. Development takes place within the coelom, and the young leave the body of the mother by way of a rupture in the anal region. A deep-sea elasipodan is known to brood its young in the ovary.

Development

Except in brooding species, development takes place externally in the sea water, and the embryo is planktonic. Development through

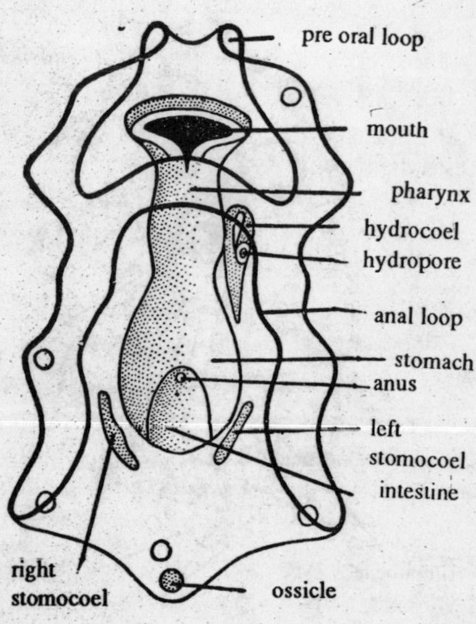

Fig. 6.7. Auricularia larva.

gastrulation is like that of asteroids. The anterior half of the archenteron separates to develop as the coelom, leaving a shorter posterior portion to become the gut. The right axohydrocoel never forms. By the third day of development, a larval stage called an auricularia has been reached. The auricularia is very similar to the bipinnaria of the asteroids, and possesses a ciliated locomotor band that conforms to the same development as the locomotor band of the bipinnaria. The *auricularia* of most species is 1 mm. or less in length, but giant 15 mm. larvae of unknown adults have been collected in plankton off Japan, New Zealand, Bermuda, and the Canary Islands. Further development leads to a barrel-shaped larva, called a *doliolaria*, in which the original cilated band has become broken up into three to five ciliated girdles

There are some species of holothuroids (Dendrochirotida) that possess a nonfeeding, barrel-sshaped vitellaria. This type of larva, which is found in crinoids and a few ophuiroids, possesses ciliated bands but no arms. *Fell* (1966) believes the vitellaria to be the most generalized type of echinoderm larva. Gradual metamorphosis, forming a young sea

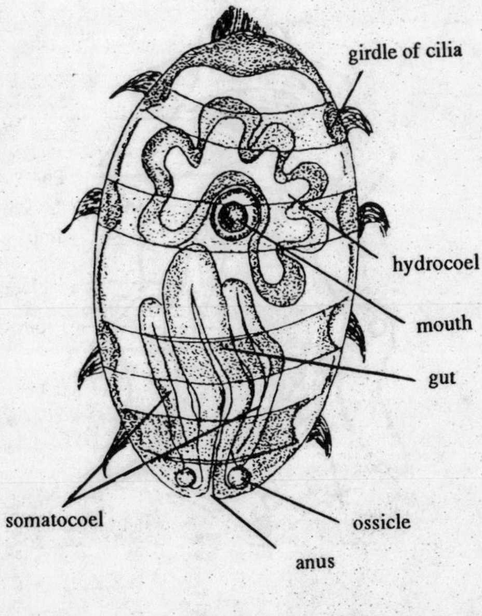

Fig. 6.8. Doliolaria larva.

cucumber but with little loss of larval features, takes place during the latter part of planktonic existence. The tentacles, which are equivalent to buccal podia, appear prior to the appearance of the functional podia. At this stage, the metamorphosing animal is sometimes called a *pentactula* larva. Eventually the young sea cucumber settles to the bottom and assumes the adult mode of existence.

Economic Importance

Dried holothurians are used as food at many places. Among South Pacific Islands and on the coasts of Queensland in Southern China these are called "beche-de-mer" or "trepang." The trade of the dried holothurians forms a good business running into thousands of dollars annually. Trepang is made of large holothurians that are called by various native names. The most favoured species is *Thelenota ananas* called by a native name meaning prickly red fish. Other species used are *Holothuria nobilis, mauritiana, scabra, lecanora, argus, edulis and echinites and Stichopus variegatus and japonicus.* The methods of preparation of holothurians to make trepang vary in different regions, but in general the cucumbers are first boiled. They throw out their viscera become short and thick. The body wall is then dried thoroughly in the sun or by smoking or by both sun and smoking. It is evident from this that trepang consists of cured body wall of certain large holothurians. Sometimes steps are taken to remove the ossicles from the body wall. People regard trepang to be highly nutritious diet constituting 35-52 per cent protein, 21-23 per cent water and 15-30 per cent ash. Carbohydrates are wanting but some fat is present. Dried material is cut into small pieces which are added to soups and stews and are said to impart a delicate flavour. When so cooked the pieces of trepang swell into a gelatinous condition and are also eaten as tidbits. In some Indo-Pacific regions the holothurians are irritated until they throw out viscera (eviscerate) and then the curverian tubules and gonads are eaten raw.

7

WATER VASCULAR SYSTEM IN ECHINODERMATA

The water vascular system or ambulacral system is peculiar to Echinodermata and not met within any other groups in the Animal Kingdom. Since the entire system is derived from the coelom the system is lined with a ciliated epithelium and filled with a watery fluid and contain corpulces. This system is used for locomotion. It probably functions in collecting and distribution of food. The ambulacral system consists of madreporite, stone canel, ring canel, radial canal, and tube feet. The system is variously modified in different classes of Echinodermata.

Class I Asteroidea.

Asteroids ambulacral system follows the general eclinoderm pattern with certain modification.

Madreporite. It is a light coloured, sieve-like, circular skeletal plate present inter-radially on the aboral surface of the disc. Its surface is marked by a number of radiating narrow, straight or slightly wavy grooves or furrows. Each furrow is furnished with minute pores which open into a sac-like ampulla through minute pose canals. The number of pores may be 250. The ampulla leads into a 's' shiped stone canal.

Stone canal. The stone canal extends downwards and opens into a ring canal, around the mouth. The walls of stone canal is supported by a series of calcaricous rings. The lumen of stone canal is lined by very tall

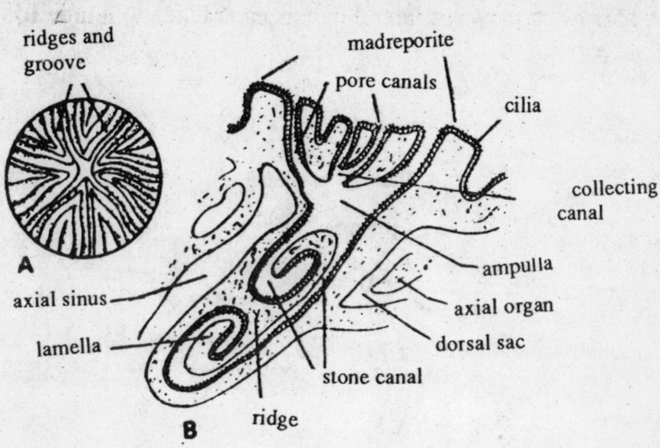

Fig. 7.1. *Asterias*. A = Madreporite seen from outside, B = V.S. through madreporite

flegellated cells. The lumenn of the stone canal is divided by a longitudinal ridge. In *Asterias* the ridge is bifurcated. In *Asteropecten*, the ridge meets the opposite wall and the interior is divided into tubes which are provided with a pair of scrolls each. *Ring canal*. It is a wide, somewhat pentagonal canal forming a ring around the oesphagus. The angles of pentagon are situated in the radial position.

Tiedmann's bodies. These are smell rounded yellow glandular sacs which open into a ring canal on its inner side, between each radius and interradius except where stone canal opens into the ring canal, these is one Tiedmann's body. Thus these are 9 Tiedman's bodies. The function of these bodies is not yet known, probably they help in the formation of amoebocytes (coelomic corpuscles).

Polian Vesicle. In most sea-stars, but not in *Asterias*, the ring canal gives off in each interradius, one, two or even four large, thin walled plan-shaped sacs known as polian vesicles. They are contractile structures and perform several functions such as formation of coelomic corpuscles, regulation of pressure etc.

Radial canal. From the outer surface of ring canal are given out 5 radial canal, one in each arm upto extremity through the ambulacral groove of the arms.

Lateral Canals. In each arm, the radial canal gives out two series of short, narrow, transverse lateral canals each leads to a tube foot. They have valves.

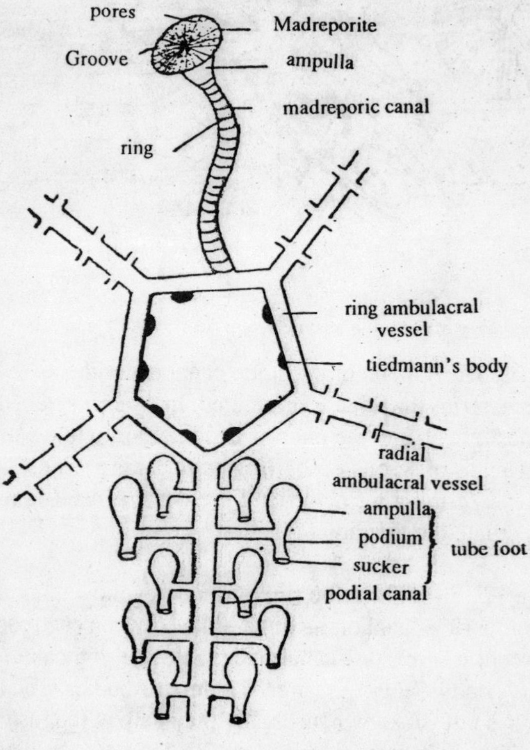

Fig.7.2. *Asterias*. Ambulacral System.

Tube feet. Each tube foot is a hollow, elastic thing walled closed cylinder. It consists an ampulla, a tube and a sucker. The tube feet are capable of greater extension and when extended they come out through ambulacral grooves.

Class - II Ophiuroidea

The water vascular system of Ophiuroidea has the same pattern like that of Asteridea. The oral shield that forms the madreporite usually bears but a single pore and canal. In addition to the oral position of the madreporite, the stone canal ascends to the ring canal. The water ring canal bears four polian vesicles. Some times polian vesicles are accompanied by long slender, tabular appendages known as *Simroth's appendages*. Tiedmann's bodies are lacking. The ring canal gives a radial canal in each radius. It descends towards the oral side where it enters the arm base, extending along the arm upto its tip and terminates into the lumen of terminal tentacle. In each ossicle, the radial canal gives rise to a pair of lateral canals, which may extend directly to the pair of ventro-lateral podia. Ampullae are absent, probably correlated with the reduction of the arm coelom. A valve is present between lateral canal and podium. The entire water vascular system is lined with a ciliated peritoneum.

Class - III Echinoidea

The water vascular system of echinoides is essentially like that of asteroides. One of the genital plates around the periproct contains pore and pore canals and functions as madreporite. A stone canal desend orally to the ring canal. The ring canal lies above the peristone is heart-urchim or just above the chewing apparatus in regular urchins and sand-dollars. The radial canal extends from the ring canal and run along the underside of ambulacral areas of the test. Five polian vesicles are given out from the ring canal. Five Tidemann's bodies are also present.

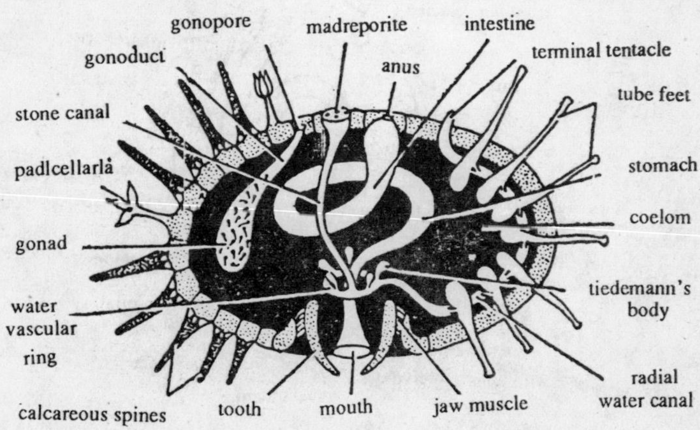

Fig. 7.3. Section of sea urchin to show water vascular system.

Each radial canal terminates in a small ampulla called as terminal tentacle. The radial canal gives off alternately on both sides lateral canals to the bases of the ampullae. The canals connecting the ampullae and podia. Each tube foot consists of an ampulla, a stalk and a sucker. The suckers are hightly developed and are provided with a system of muscles and supporting ossicles.

Class - IV Holothuroidea

Although he ambulacral system is basically like that of other echinoderms but the madreporite in most species have lost connection with the body surface and in being unattached in the coelom. Pore and pore canals are present in madreporite. The madreporite hangs just beneath the base of pharynx and is connected to the ring canal. The ring canal encircle the pharynx and gives rise polian vesicles. The number of polian vesicles varies considerable, 1 in *Cucumaria,* 3-4 in *Thyone* or as many as 10 or even 50 in certain Apodida. From ring canal 5 radial canals

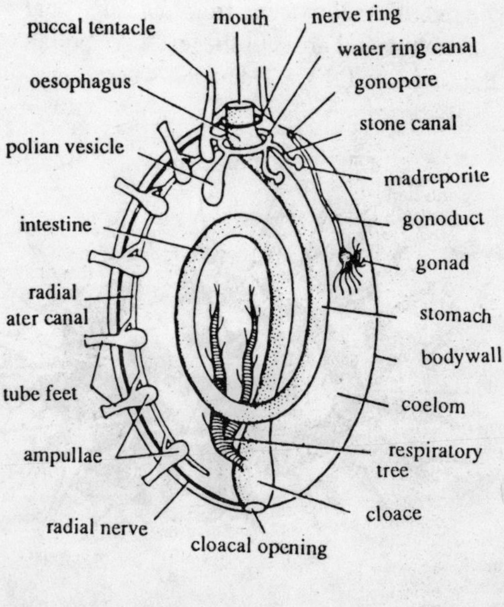

Fig. 7.4. Section of *Holothuria to* show water vascular system.

pass upward to the inner side of the calcereous ring. Each radial canal gives off small canals to the tentacles. The radial canal reduces in diameter and after passing through a hole in circum-oesophageal caleare-ous ring, it runs backwards along the inner surface of ambulacral area. Here lateral canals supply the podia and their ampullae.

In the Anodida, the tube feet are absent hence the water vascular system is limited to the oral ring vessel, the polian vesicles and buccal tentacles.

Class - V Crinoidea

The madreporite is absent in Crinoides. The ring canal encircles the mouth gives off at each inter-radius a large number of stone canal (30-50)

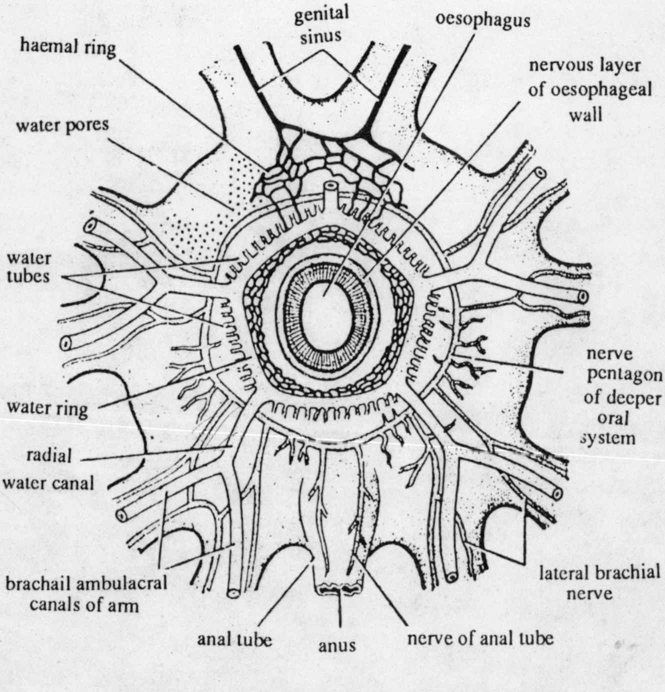

Fig.7.5. *Antedon*. Showing water vascular System.

which open into coelom. At each radus of ring canal, a radial canal extends into each arm just beneath the ambulacral groove and forks into all of the branches and into the pinnules. From the radial canals extend lateral canals supply the podia. There are no ampullae and one lateral canal supplies the group of three podia except in buccal region. Peculiar to crinoids are 500-1500 minute ciliated canals also called *ciliated funnels*. These funnels perforate the wall of tegmen and open into the underlying coelomic spaces. These openings compensate for the absence of madreporite, permitting sea water to enter the coelom and maintain a proper fluid pressure in the body and the water vascular system.

8

LARVAL FORMS IN ECHINODERMS

In echinoderms although the sexes are separate but there is no sexual dimorphism. The gametes are released directly in sea water where fertilization occurs. The development may be direct or indirect. In the direct development there is no larval stage but in the indirect development different kinds of free swimming larval forms are found. In each class, a few members, chiefly polar or deep - sea species, are viviparous. They rear the young in the brood opuch-like structure found in their body.

The most striking feature of the larva is its bilateral symmetry, which is in marked contrast to the radial symmetry of the adult, After a free-swimming life, the bilateral larva undergoes metamorphosis, in which the radial symmetry of adult is developed.

Class 1 Asteroidea

Bipinnaria Larva. The egg hatches into a larva which develops the cilia and starts swimming freely. The alimentary canal is formed and larva feeds upon diatoms. In the wall of the gut, powerful ciliary band is present which helps in ingestion. Later the cilia of the body become restricted to the two cilary bands characteristic of Eleutherozoan larvae. The two lateral bands connect in front of the mouth as a *pre-oral loop* and infront of the anus as a *pre-anal loop*. The preoral loop later separates or in some cases develop independently into an anterior ciliated around the body. Three lateral lobes are formed on either side of the larva. The larva at this stage is called bipinnaria larva. Its development is completed between 2 - 7 days.

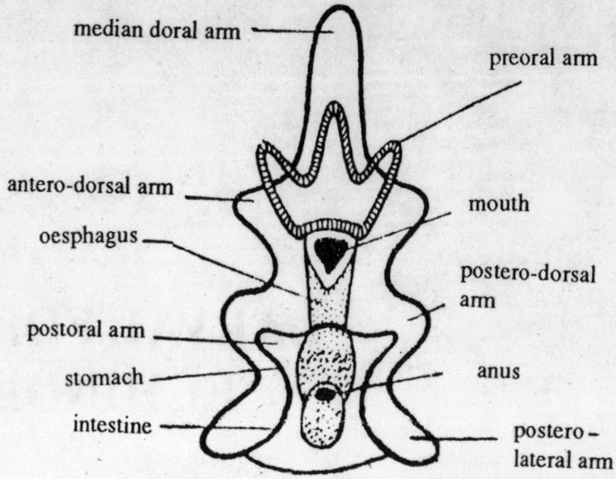

Fig. 8.1. Bipinnaria larva.

Internal development of bipinnaria. The tip of archenteron gives rise to the mesenchyme. The archenteron gives rise to a coelomic sac, the hydroenterocoel, on each side. The two occupy the right and left of the archenteron and gradually elongates anteriorly in the plan of the long axis of the larva to form U - shaped coelom. The posterior ends of the lateral pouches become cut off from anterior *axocoels* and hydrocoels as right and left *somatocoels.* The left hydorcoel connects with the dorsal surface to form the *hydropore* without ectodermal invagination. On the ventral surface, the invaginated part of the ectoderm is connected with the endoderm. The alimentary canal of the larva can be divided into mouth, oesophagus, stomach, intestine and anus. The blastopore forms anus. At the time of metamorphosis, the right somatocoel and axohydro coel are reduced while the left axocoel gives rise to water ring and radial canals. The axocoel separates from hydrocoel and contributes to stone canal. The madreporite is either formed by the rearrangement of of the mesenchyme cells or by invaginated part of right axohydrocoel. After swimming for a few weeks the bipinnariea larva changes to next larval stage, the *brachiolaria.*

Brachiolaria larva. At the time of the changing of the bipinnaria into the brachiolaria, there additional arms are formed in the preoral lobe. These are called *brachiolar arms* of these one is median and two are laterals. These arms contain coelomic extensions and adhesive cells at their tips. An adhesive glandular region is found at their bases. It acts like a sucker. The appearance of sucker marks the beginning of metamorphosis.

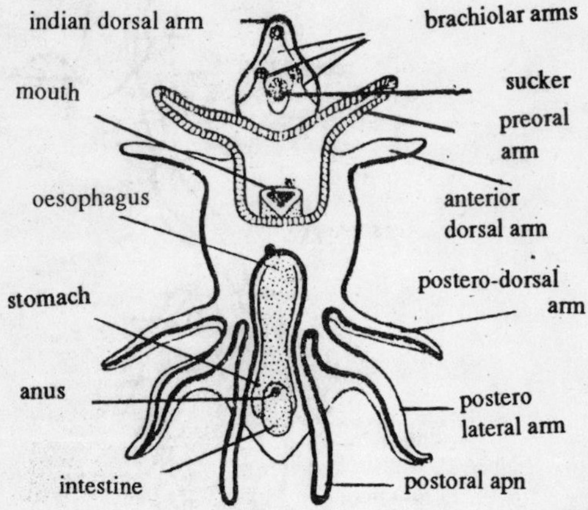

Fig. 8.2. *Branchiolaria larva*.

Metamorphosis. The bilaterally symetrical larva gradually metamorphoses into the radially symmetrical adult. After about 6 or 7 weeks the brachiolaria settles down on some solid object or to the bottom where it remains temporarily attached by its fixing processes. The anterior end of the larva degenerates and forms only an attachment stalk, and the adult body develops from the posterior end of the larva. The left side becomes end of the larva. The left side becomes the oral surface and right side aboral. The adult arms appear as extensions of the body. Internally the entire digestive tract degenerate. All these part formed anew and in a position coinciding with the adult radial symmetry. The somatocoel forms the major part of coelom. The left axohydorcoal forms the water vascular system and from the hydrocoels develop five pairs of projections, two in each of the developing arms. These projections represent the cavity and coelomic lining of first pair of podia in each arm. As about this time, sea star is not more than 1 mm in diameter.

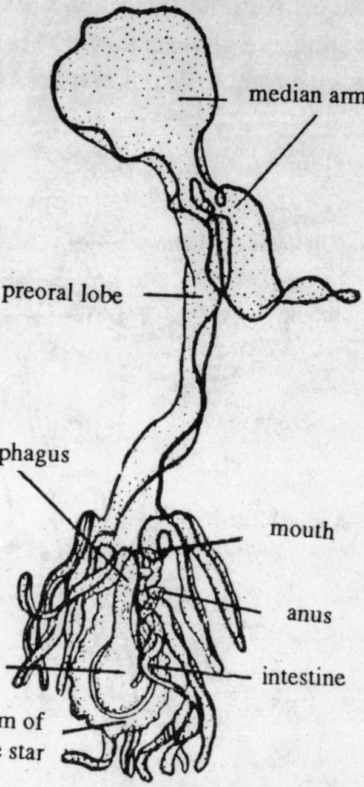

Fig. 8.3. Larva showing *metamorphosis*.

In some species, *Astropecten,* the development is shortened. Here the brachiolaria larva disappears and the bipiunaria larva metamorphoses into the adult after about three months. In *Asterina gibbosa,* the bipinnaria larva is not formed, the larva develops an adhesive apparatus, homologous to the brachiolar arms and sucker, which metamorphoses in adult. In *Luidia,* a giant and peculiar larva is formed known as *bipennaria asterigera.*

Class 2 Echinoidea

Echinopluteus. In Echinoidea, a free swimming, microscopic *echinopluteus larva* is formed after gastrulation in about 7 - 30 days. The gastrula becomes somewhat cone-shaped and gradually develops into echinopluteus. The invaginated part of stomodaeum becomes connected with the archenteron and the gut is differentiated into mouth, oesophagus, stomach and intestine. The blastopore forms the anus. It bears six pairs of arms called *preoral, anterio-lateral, anterio-dorsal post-oral, poster-odorsal* and postero-lateral. The *poster-lateral* arms are very short and directed outwards or backwards. In some cases, the anterio-dorsal arms may not develop. Thus there may be 5 or 4 pairs of arms in place of six pairs. The ends of these arms are pigmented and the skeleton is made up of calcareous rods, present in arms. These rods are originate from spicules secreted by mesenchyme. These rods may be thorny or simple or fenestrated or branched. The locomotion is performed by ciliated bands. In some cased the bands become thickened and known as *epauletes*. In *Arbacia* and Cidaris, the ciliated lobes are formed between the bases of the arms which are known as *vibratile lobes auricular lobes* or *auricles*.

The archenteron gives off hydroentesocoels which continue to axocoel, hydrocoels, and somatocoels. The ectoderm forms and invagination on the left side which enlarges and forms a vestibule. The hydorcoel gives off five *radial canals* and give primary podia. The lantern of Aristotle is formed from left somatocoel.

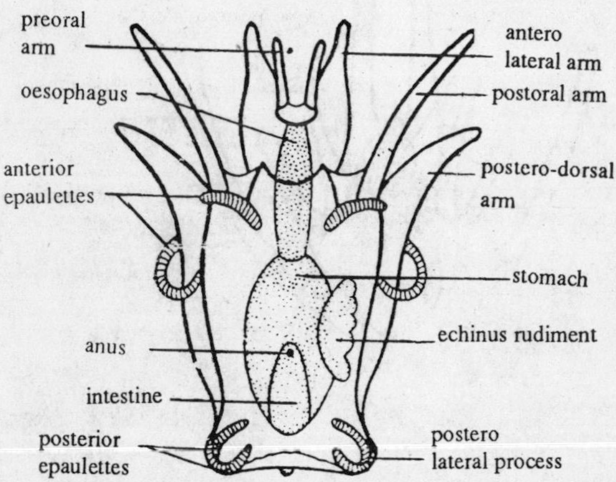

Fig. 8.4. Echinopluteus larva

Metamorphosis. The echinopluteus gradually sinks to the bottom. There is no attachment as in asteroids. Metamorphosis is rapid, taking place in about an hour. Internal development is basically like that of asteroids, although differing in detail. Young urchins are no larger then 1 mm.

Class - 3 Ophiuroidea.

Ophiopluteus. The pluteus larva is also formed in the brittle stars and it is called ophiopluteus. It is similar to echinopluteus but the number of arms is less in ophiopluteus. Its posterio-lateral arms are formed first. The poserio-lateral, postoral and posterodorsal arms are developed respectively on 4th, 10th and 18th days. The ciliated bands accompany the arms edges. The coelomic chamber and archenteson are found in larva.

The internal development follows the same pattern as in other classes.

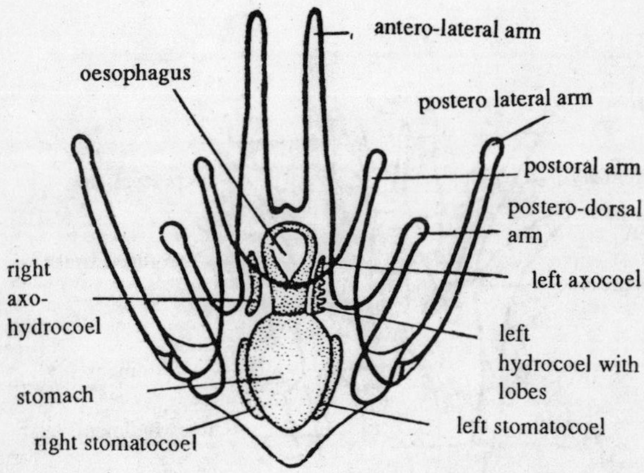

Fig. 8.5. Ophipluteus larva.

Metamorphosis. Metamorphosis starts after the appeasenal of the skeleton in larva. Metamorphosis takes place in free-swimming stage. In viviparous species like *Amphira viivpetra,* the pluteus stage is absent.

Class - 4 Holothuroidea

Auricularia. After the gastraulation and formation of coelomic sacs and gut, within three days a free swimming auricularia larva is formed. It is transparent and palegic in nature. It measures .5 to 1 mm in length. It swims with the help of ciliated bands. One of these bands encircling the mouth and is called *pre-oral loop.* Similarly, the ciliated band around the anal opening is called the *anal loop*. Internally, the larva contains a curved gut. The stomach is *sacciform.* Hydrocoel and the right and left somatocoels are also present. Lobes appear in the hydrocoel forming the primary tentacles which are connected with the hydropore by means of a duct. Some giant auricularia of unknown adults measuring 15 mm in length, have been reported from Bermuda, Japan and Canary islands. They possess a frilly flegellated band.

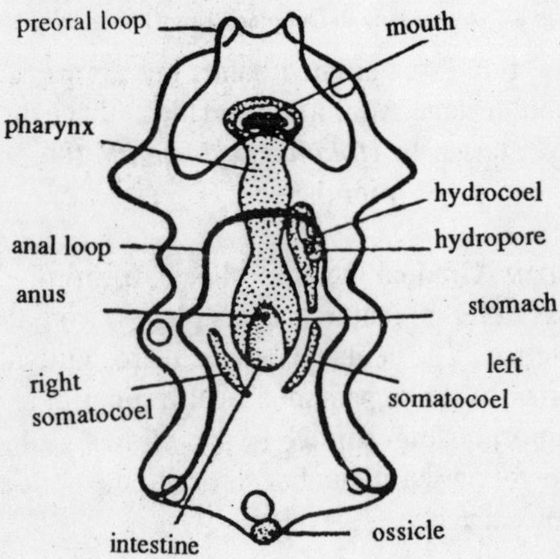

Fig. 8.6. Auricularia larva.

Doliolaria. The auricularia larva soon changes into a barrel-shaped form, the *doliolaria larva.* In doliolaria stage, the continuous ciliated bands break to form 3-5 flegellated spheres. The mouth has shifted to the anterior pole and the anus to the posterior pole.

These are some species of holothuroids that possess a non-feeding barrel-shaped vitellaria. This type of larva, which is found in crinoid and

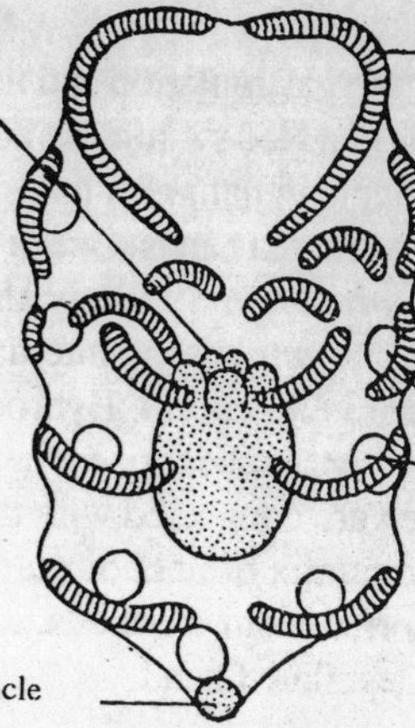

Fig. 8.7. Auricularia to Doliolaria (transitional stage).

a few ophuiroids, possesses ciliated bands but no arms. In *Cucumeria planchi*, the auricular stage is ommited and the embryo directly develops into the doliolaria larva. In *Holothuria floridana* the embryo directly develops into young stage from the egg.

Metamorphosis. Gradual metamorphosis, forming a young animal but with little loss of larval features, takes place during the latter part of planktonic existence. The tentacles, which are equivalent to buccal podia, appear prior to the appearance of the functional podia. At this stage, the metamorphosing animal is sometimes called *pentaculata larva*. Finally the young sea cucumber settles to the bottom and assumes the adult mode of existence.

Class - 5 - Crinoidea

Doliolaria. After gastrulation and embryonic development of coelom and gut, the larva hatches as doliolaria larva. It is free swimming. An apical sensory plate, possessing a bunch of cilia is present at its anterior end 4 or 5 ciliated bands are also present on its body. Near the apical plate in mid-ventral line the larva has an adhesive pit over the first ciliary band. The *stomodeum* or *vestibule* is present between 2nd and 3rd ciliated band. The skeleton also develops at this stage. The larva attaches itself to some object after the development of the prospective organs of the coelomic sac and its internal organs of the coelomic sac and its internal organs rotate at an angle of 90° from the ventral to the posterior direction. The larva forms a stalk and at this stage it is known as the the *cystidean* or *pentacrinoid larva.* After some time, it metamorphoses into an adult.

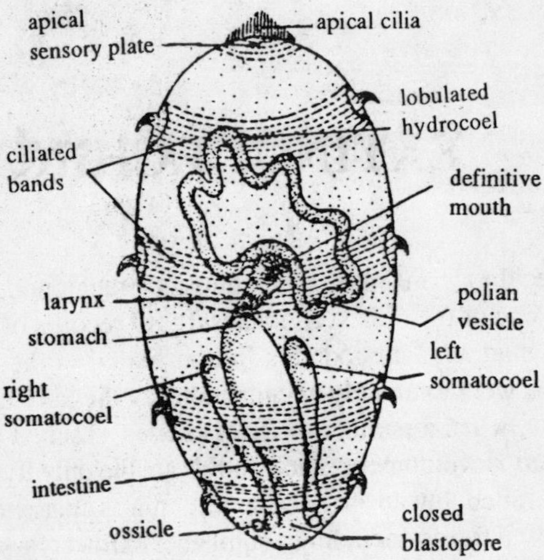

Fig. 8.8. Doliolaria larva.

Significance

All the larval forms of echinoderms have a bilateral symmetry. Hence it is believed that the ancestor of enchinoderms was a bilaterally symmetrical animal. According to *Semon* (1888) the radial symmetry but this secondary as the primary radial symmetry is seen in coelenterates and porifera. The adult echinoderms are more primitive than their larvae because the adults possess the characters of the lower animals like coelenterates etc. During metamorphosis the advanced larva becomes a primitted adult. Such a metamorphosis is called retrogressive metamorphosis.

9

EXTINCT ECHINODERMS

The echinoderms rank with mollusks, brachiopods, and arthropods in having one of the richest and oldest fossil records of any group in the Animal Kingdom. Echinoderms first appeared in the lower Cambrian period and were extremely abundant during the later periods of the Paleozoic era, when a number of fossil classes reached the peak of their evolutionary development. The crinoids are the only living echinoderms that are attached, but an attahced condition was characteristic of the majority of fossil forms, including a number of extinct classes. These groups are particulary interesting in that they display many primitive features of the phylum.

Like living crinoids, most fossil forms were probaby suspension feeder in which the podia and upward-directed ambulacral grooves were food-catching and food-conducting structures. In most attached fossil species, the pepripheral end of the groove was extended upon a slender pinnule-like projection (the bachiole). The term *brachiole* is used to distinguish these body extensions from the heavier arms of crinoids and stelleroids. Brachioles varied in number from a few to hundreds and like the arms and pinnules of crinoids represented an adaptation that increased the surface area for food collection. Since these fossil forms were sessile, and since the oral surface was always directed upwards, the original function of the podia could not have been locomotor. The skeletal system of the extinct attached echinoderms was somewhat like that of the crinoids, with the exception that the ossicles (plates) of the crown were not limited to the aboral surface (calyx) but extended orally to the mouth, thus enclosing the internal organs within a test (or theca). Protective cover plates, which could fold over the ambulacral grooves and mouth, were generally present. The anus was usually located eccentrically in one of the interradii, as in the crinoids.

The oldest group of extinct echinoderms is the Eocrinoidea, known from the lower Cambrian to the Ordovician. Eocrinoids were stalked or stalkless echinoderms with an enclosed theca as in *Macrocystella.* At the upper, or oral, end were five ambulacra and five to many brachioles. The pentamerous and oral position of ambulacra and branchioles gives these animals a superficial resemblance to crinoids, although they are actually more similar to cystoids. The class Cystidea was a group of attached species that ranged from the middle Ordovician period to the Permian. The theca of cystoids was more or less oval, with the oral end directed

Fig. 9.1. *Macrocystella.*

upward and the aboral end attached to the substratum directly or sometimes by a stem. A characterisstic feature of the cystoids was a system of pores that perforated the theca. The pores were either dispersed over the body surface (diplopores) or confined to special areas (rhombopores). In either case, the pores wer probably part of a system for gas exchange permitting either an outward flow of coelomic fluid or an inward flow of sea water. The three, or more commonly five, ambulacra were radially arranged and extended outward and downward to varying degrees over the sides of the theca, in some species to the aboral pole. Small brachioles were either located around the mouth or mounted on the plates along each side of the ambulacral grooves. Branches from the groove extended up each brachiole.

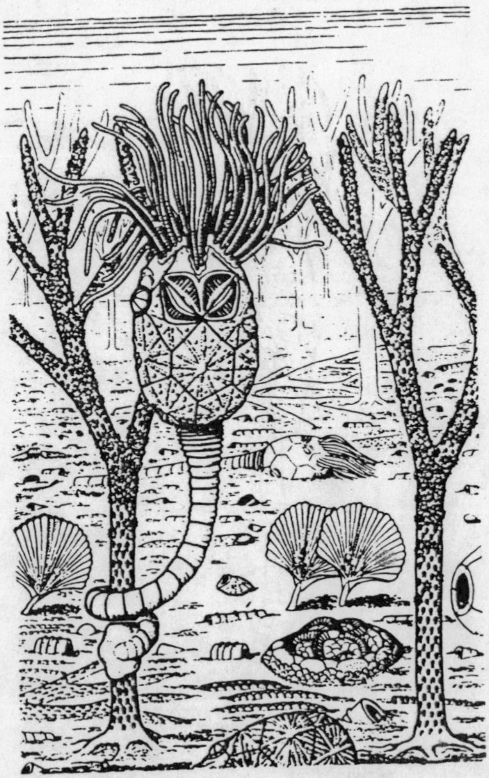

Fig. 9.2. *Lepadocytis*

The class Blastoidea is another group of extinct attached echinoderms that flourished in the Paleozoic seas. The blastoids first appeared in the Ordovician period, reached their peak in the Mississippian period, and became extinct during the Permian period. *Pentremites* is a very common blastoid fossil found in Mississippian limestone. Like cystoids, blastoids possessed an oval theca that was attached aborally by a short stalk or was attached directly to the substratum. The arrangement of thecal plates was simpler and more regular than in cystoids. Five ambulacra extended from the mouth at the oral pole down the sides of the theca and frequently were elevated as broad ridges that alternated with the depressed interambulacral areas. The margins of the ambulacra were bordered by a single row of slender, closely placed brachioles, onto which extended branches from the ambulacral groove. In oral view, the brachioles gave the appearance of a thick fringe along the border of a five-pointed star.

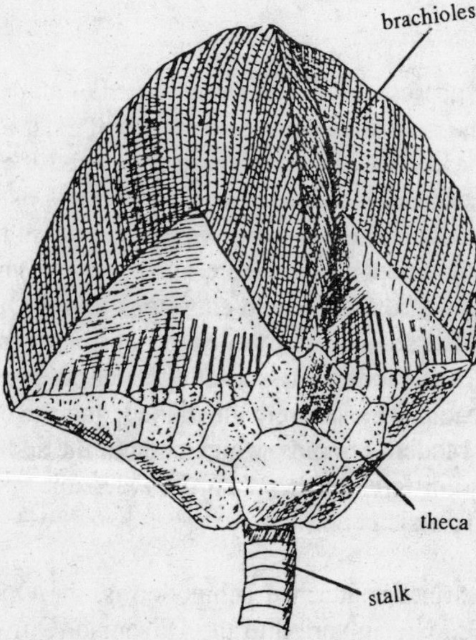

Fig. 9.3. *Blastoidocrinus*.

Peculiar to blastoids was a system of folds, or folds and pores, called *hydrospires,* located at each side of the ambulacra. The hydrospires are thought to represent a gas exchange mechanism by which water circulated inwardly through the thecal plates, allowing greater surface area for the exchange of gases between the coelom and the sea water. The hydrospires were thus somewhat like reversed asteroid papulae, the folds projecting into the coelom instead of externally from the coelom. The Edrioasteroidea is a class of extinct attached echinoderms that first appeared in the lower Cambrian period and ranged into the Pennsylvanian period. The theca was oval or discoid and composed of imbricated, abutting, or fused plates. The presence of tubercles on the plates indicates that some species possessed spines. There is little to indicate that other extinct attached echinoderms had spines. Some species had stalks but the discoid species were stalkless and lived attached to shells or other objects. There were no brachioles present, and the five ambulacra, which extended over the theca, were straight or curved. Some of these animals thus looked like a brittle star wrapped around a ball. Of particular interest was the presence of pores between the ambulacral plates through which the podia extended. Movable cover plates protected the podia and food grooves.

The crinoids possess the richest fossil record of all the echinoderm classes and are the only attached species that still exist today. Typical crinoids first appeared in the Ordovician period, and they left a fossil record in every succeeding period. The class reached its climax in the Mississippian period, although there was a second somewhat lesser evolutionary development during the Permian period. During these periods, certain shallow seas supported enormous fauna. During the Permian period, such a sea covered the island of Timor, and the rocks composing this island today are one of the richest sources of fossil specimens. Paleozoic crinoids were all stalked, and the earlier species lacked pinnules. Modern crinoids, which contain the stalkless comatulids and belong to different orders than the Paleozoic species, did not appear until the Triassic period.

A class of extinct unattached echinoderms, the Ophiocistioidea, lived from the Ordovician period to the Devonian. Only a few dozen fossil species are known. The body was encased in a test that was flattened orally and was domeshaped aborally. Except in *Volchovia,* there was no anus and the madreporite was located eccentrically on the oral surface. The mouth, which was surrounded by a peristome, was located in the center of the oral surface. A chewing apparatus composed of five jaw like plates was also present. There was apparently only one gonad, but multiple gonopores opened in the same interambulacrum as the madreporite. Five radially arranged ambulacra extended peripherally from the peristome but were limited to the oral surface. Most remarkable were the three pairs of giant podia that occupied the ambulacra. Each podium was covered by small overlapping scales and was thus preserved with the test. There seem to have been pores present between the podial plates from which papula-like structures could have emerged.

As would be expected from the nature of the skeleton, the holothuroids have the poorest fosil record and the echinoids have the best fossil record of the living classes of unattached echinoderms, but compared with attached echinoderms, but compared with attached forms the fossil record of unattached species is sparse. Although fossil asteriods and echinoids first appeared in the Ordovician period and ophiuroids in the Mississipian, Paleozoic species are relatively rare, these classes are much better represented in Mesozonic and Cenozonic rocks. The earlier echinoids, including a number of extinct orders, were all regular forms, and the first fossil heart urchins and sand dollars appeared in the Triassic period. Rich deposits of fossil echinoids are knon, but the fossil distribution of this class is too localized for them to be of any value as index fossils.

10

PHYLOGENY OF ECHINODERMATA

The classes of echinoderms, both living and extinct, are placed into to subphyla, the Pelmatozoa and the Eleutherozoa. The Pelmatozoa contained attached species having the oral surface directed upard, the Eleutherozoa contained unattached echinoderms having the oral surface directed donward. It is now recognized that such a division is an artificial one from the standpoint of phylogenetic relationships, and these subphyla have been abandoned. However, the terms pelmatozoan and eleutherozoan have been retained to designate the relationship of echinoderms to the substratum. Thus, crinoids are siad to be a pelmatozoan group, and sea stars are siad to be eleutherozoan.

The origin of the echinoderms and the phylogenetic relationships of the subphyla continue to be unsolved questions and subjects of much speculation. Despite the extensive fossil record, paleontological evidence is still insufficient in many important areas. Before revieing some of the older ideas that are still current, as ell as some more recent views, it may be helpful to enumerate a few reasonable assumptions that would be accepted by many zoologists.

1. Echinoderms evolved from motile, bilaterally symmetrical ancestors that possessed a tripartite coelom.
2. As is true of other animals, the echinoderm skeleton and radial

symmetry probabaly represent adaptations to a sessile existence, at least initially.

3. The original function of the water-vascular system was feeding and not locomotion.

A classical theory, and one that would still find many supporters, holds that from a bilateral tricoelomate free-moving ancestor, some group became attached to the bottom and assumed a sessile mode of life. As in many other animal groups, such a sessile existence resulted in a shift to a more adaptive radial symmetry. Attachment apparently took place at the anterior end of the animal. Based on the metamorphosis of living echinoderms, the change in symmetry involved a clockwise 90^0 rotation of the animal so that the left side became the upper (oral) side and the right side became the lower (aboral) side. Simultaneously, the mouth moved around to the original left side, now the upper side. The two right anterior coelomic sacs (axohydrocoel) were reduced, and the two left sacs became the water-vascular system, which functioned in suspension feeding.

During the course of these changes, the echinoderm skeleton probabaly evolved as a supportive and protective structure for the sessile animal. *Nichols* (1967) suggests that the pentamerous form of radial symmetry arose in conjunction with the skeleton. The suture planes - the junction point between two skeletal plates - represent a weak point in the body wall from a structural standpoint, and it would have been advantageous to the animal not to have had two such suture planes opposite one another. This advantage could be attained only by an odd number of ossicles forming the circumference of the body wall. The smallest number would be five if the animal were to be truly radial. At this point, we have a suspension-feeding echinoderm that was both sessile and radial. This stage in the evolution of echinoderm symmetry is illustrated by the extinct and living Crinozoa.

After attaining a radial symmetry, some of these sessile echinoderms became detached and reassumed a free-moving existence. The radial symmetry was retained, but the oral surface, which was directed upward in sessile forms, was placed against the substratum; the aboral surface became the functional upper side of the animal, and the water-vascualr system was utilized in locomotion. Sea stars, brittle stars, and sea urchins all illustrate such a free-moving, radial existence.

This theory is supported by the extinct and living crinozoan fauna and by embryological evidence, such as the attached metamorphosis of asteroids. However, there are also some objections to this theory. The oldest echinoderms, all from the early Cambrian, are three very diverse groups: the eocrinoids, the edrioasteriods, and the carpoids. The pelmatozoan habit of the eocrinoids and the edrioasteroids is compatible with the idea that the first echinoderms were radially symmetrical and attached animals, but the irregular carpoids are not. Either the carpoid symmetry is secondary, in which case the earliest echinoderms were not pentamerous, radially symmetrical animals.

A second principal problem with this classical theory of echinoderm evolution is the lack of any forms that bridge the great gap between the pelmatozoan and eleutherozoan habits. The echinozoan edrioasteriods have frequently been suggested as a bridge group, for they were pelmatozoan and some were unattached. However, it is difficult to visualize how the oral surface could have come to be directed downward with sufficient intermediate stages to permit the necessary functional adjustments.

Fell (1965) postulates that the eleutherozoan habit and not the pelmatozoan one was the most primitive mode of existence in echinoderms, and that the sessile pelmatozoan habit arose secondarily, a number of times independently, within the phylum. The living echinoids and most of the holothuroids are believed to be decended from forms that were always free-moving. The helicoplacoids are regarded as being closest to the ancesteral echinoderms, and the earliest echinoderms, and the earliest echinoderms as perhpas being burrowing animals. Fell's views have the advantage of reconciling the conflicting pelmatozoan and eleutherozoan modes of existence within the phylum, but they do not explain the original adaptive significance of the echinoderm skeleton, which is a basic feature of the phylum, nor do they explain the pentamerous radial symmetry of such eleutherozoan forms as the echinoids.

Embryology sheds little light on the interrelationships of living echinoid pluteus larvae was formerly thought to indicate an evolutionary relationship. However, many echinologists now believe that there is considerable convergence in the larval forms of different classes.

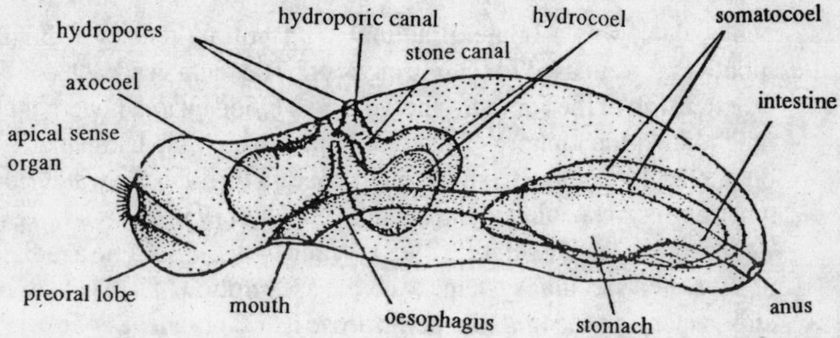

Fig. 10.1. Hypothetical dipleurula ancestor.

There has also been much speculation aboiut the nature of the ancestral pre-echinoderms. A classical view has been that of *Bather* (1900), who postulated that echinoderms evolved from a creeping, soft-bodies, bilateral animal (called a dipleurula ancestor), which possessed thrree pairs of coelomic sacs. The water-vascular system was a new development of the coelom in the evolution of the phylum.

Another theory, originally proposed by *Semon* (1988) and *Bury* (1895), has been supported by *Hyman* (1955). According to these zoologists, the common ancestor of echinoderms, which they call a

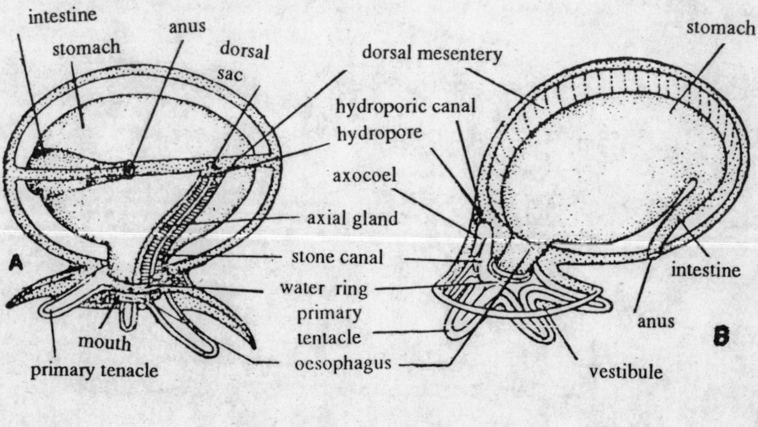

Fig.10.2. Pentactula ancestor.

pentaculata, was a bilateral animal with hollow tentacles aroudn the mouth that were used for capturing food. The tentacles, each containing an extension of the coelom, a nerve, and a blood channel, were thus very much like a lophophore. The coelom into which the tentacular coelum opened had separated from the main coelom forming the water ring, and it had a special opening to the outside, the *hydropore* and the *stone canal*. According to this *Pentactula Theory,* each tentacle became a radial canal of the water-vascular system. With the assumption of a sessile existence and suspension feeding, the podia arose as side branches of the original tentacles. The arms (brachioles) developed with the addition of protective plates in the tentacle walls, and the ambulacral groove evolved ask a means for conducting captured food particles to the mouth.

The Pentactula Theory is attractive in that it provides a functional precursor of the water-vascular system and also links the echinoderms and the lophoporates. *Nichols* (1967) has suggested the Sipuncula, with their circle of hollow oral tentacles, as ancestors of both the lophophorates and the echinoderms.